Three-dimensional structure of wood

SYRACUSE WOOD SCIENCE SERIES, 2
WILFRED A. CÔTÉ, *editor*

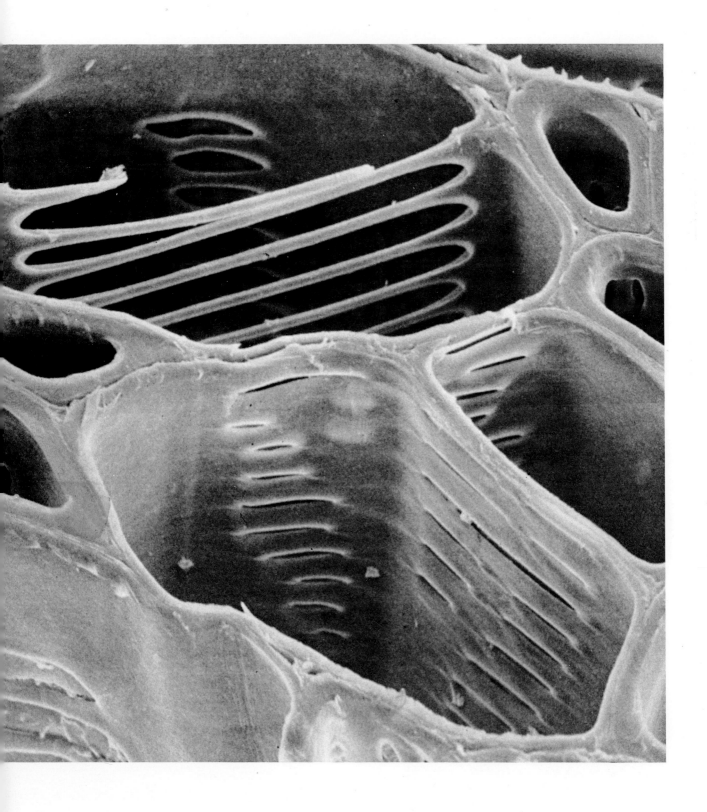

Frontispiece. *Vessel structure in* Laurelia novae-zelandiae

Three-dimensional structure of wood

A SCANNING ELECTRON MICROSCOPE STUDY

B. A. MEYLAN
Physics and Engineering Laboratory
Department of Scientific and Industrial Research
Lower Hutt, New Zealand

and

B. G. BUTTERFIELD
Botany Department
University of Canterbury
Christchurch, New Zealand

KLINCK MEMORIAL LIBRARY
Concordia Teachers College
River Forest, Illinois 60305

SYRACUSE UNIVERSITY PRESS

Contents

Preface	5
The study of wood	6
Materials and methods	7
The cell wall	8
Pits	12
Distribution and patterns of wall pitting	22
Other wall sculpturing	28
Perforation plates	33
The structure of rays	39
Axial parenchyma	46
Growth rings	49
Vessel distribution	51
Resin canals	58
Tyloses	61
Gymnosperm wood	64
Angiosperm wood	67
Reaction wood	73
Index	80

First published in the U.S.A. 1972 by Syracuse University Press, University Station, Syracuse, New York 13210

Printed in Hong Kong

ISBN 0-8156-5030-2

© 1972 B. A. Meylan and B. G. Butterfield

All rights reserved. No part of this publication may be produced, stored in a retrieval system or transmitted in any form or by any means, electronic, mechanical, photocopying, recording, or otherwise, without the prior written permission of the publisher.

This book is sold subject to the condition that it shall not, by way of trade or otherwise, be lent, re-sold, hired out, or otherwise circulated without the publisher's prior consent in any form of binding or cover other than that in which it is published and without a similar condition including this condition being imposed on the subsequent purchaser.

Preface

This book is a collection of scanning electron microscope photographs selected to illustrate various features of the structure of wood. Although it is intended to supplement general plant anatomy texts for Botany and Forestry students at the undergraduate level, it should also be a useful work for anyone interested in wood science. The lists of further reading included in each section contain only general texts and the more recent and relevant papers on each topic.

We would like to express our thanks to Mr R. R. Exley for technical assistance in the preparation of the wood samples, Mrs K. A. Card and Mr L. A. Adamson for assistance with the photography, and Miss E. Soper and Mrs T. Arkesteyn for typing the script. This project was carried out on the Cambridge scanning electron microscope of the Physics and Engineering Laboratory, Department of Scientific and Industrial Research, Lower Hutt. We would thank Mr W. S. Bertaud and Dr N. E. Flower of the electron microscope laboratory for their co-operation and Dr M. C. Probine, the Director of the Physics and Engineering Laboratory and Professor W. R. Philipson of the University of Canterbury for their encouragement throughout the project.

B.A.M.
B.G.B.

February 1971

The study of wood

The light microscope and its modifications – polarizing, and interference microscopes, have been used extensively in the study of the structure of wood. By cutting thin sections, normally about 10-15 μm thick in the transverse, radial longitudinal and tangential longitudinal planes (fig. 1), it has been possible to obtain a fairly complete picture of the gross structure of wood. The preparation of macerations has supplemented this information, enabling the component cells to be seen separately.

The advent of the transmission electron microscope has added considerably to our knowledge of the anatomical features of wood at the ultra-structural level. By using ultra-thin section or carbon replicas this instrument has enabled us to see structures whose dimensions are below the resolving power of the light microscope. Because of their small depth of field, however, both the light microscope and the transmission electron microscope illustrate material in only one plane at a time. That is, they show us only a two dimensional picture of wood. In order to construct a three dimensional model of any wood feature it is necessary to prepare serial sections of the material and construct a three dimensional diagram by hand drawings. This is not only time consuming, but also requires a certain ability on the part of the wood anatomist to picture in his mind the three dimensional nature of the material.

The very great depth of field obtainable on the scanning electron microscope has rendered it a powerful tool in wood science. This instrument, though not capable of the very high magnifications possible on the transmission electron microscope, has a depth of field 300 times greater than that obtainable from a light microscope and is capable of resolution down to 20 nm. This enables us to study the structure of wood in virtually three dimensions (fig. 2) over a considerable magnification range.

Further Reading

COLLETT, B. M. 1970. 'Scanning electron microscopy: A review and report of research in wood science.' *Wood and Fibre,* **2**, 113-133.
ECHLIN, P. 1968. 'The use of the scanning reflection microscope in the study of plant and microbial material.' *Journal of the Royal Microscopical Society,* **88**, 407-418.
FINDLAY, G. W. D. and LEVY, J. F. 1968. 'Scanning electron microscopy as·an aid to the study of wood anatomy and decay.' *Journal of the Institute of Wood Science,* **4**, 57-63.
SCURFIELD, G. and SILVA, S. R. 1969. 'Scanning electron microscopy applied to the study of the structure and properties of wood.' *Proceedings of the Second Annual Scanning Electron Microscope Symposium,* 185-196.
THORNTON, P. R. 1968. *Scanning Electron Microscopy.* Chapman and Hall, London.

Materials and methods

The photographs used in this text were obtained on a Cambridge Series II scanning electron microscope. Cubes of wood about 3-4 mm per side were cut from air-dried blocks of various trees. These were first softened by boiling in water before the final surface cuts were made by hand using a new razor blade for each surface. The cubes were then mounted on standard stubs, transferred to a high vacuum evaporating unit and lightly coated first with carbon and then with approximately 40 nm of gold palladium while being rotated at about 150 rev/min. The specimens were then examined in a vacuum dry state in the column of the microscope.

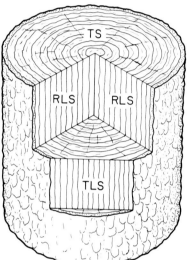

Figure 1. *This line diagram illustrates the three principal planes in which wood is normally viewed, or sections cut, in order to study its structure.* **TS,** *transverse section;* **TLS,** *tangential longitudinal section; and* **RLS,** *radial longitudinal section. Tangential sections are truly tangential over only small areas, the area becoming more significant the larger the diameter of the stem being examined.*

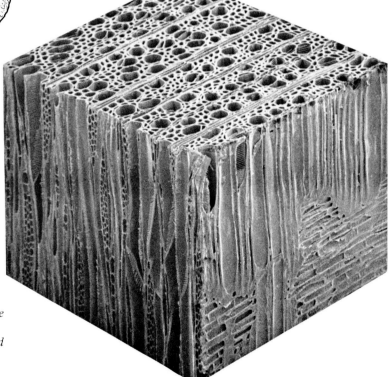

Figure 2. *A block of wood cut from* Laurelia novae-zelandiae *A. Cunn. showing the transverse plane to the top, the tangential longitudinal plane to the left and the radial longitudinal plane to the right.* (× 68)

The cell wall

Plant cell walls are built up of cellulose microfibrils variously orientated (fig. 3), and incrusted with a number of other compounds. They are subdivided into *primary walls* and *secondary walls* according to their time of formation. The primary wall develops first and is often stretched during the differentiation of the cell. It is the only wall found in some cells. The secondary wall is laid down on the inside of the primary wall, usually after elongation of the cell has ceased, and is a characteristic feature of almost all wood cells.

The primary or outer wall of a wood cell consists of a thin network of cellulose microfibrils, irregularly and loosely arranged and incrusted with hemi-cellulose, lignin, and other compounds. The secondary wall is laid down inside the primary wall and in most xylem cells is considerably the thicker of the two (fig. 4). It can be subdivided into three layers according to the orientation of the microfibrils within it. The layer nearest the primary wall is termed the S1 layer and the microfibrils in it are orientated nearly perpendicular to the long axis of the cell. The middle, or S2 layer is by far the thickest and is built up of microfibrils running at a small angle to the long axis of the cell. The S3 layer, lying nearest to the cell lumen, is a thin layer with the microfibrils again orientated in a nearly transverse direction. There may be a gradual transition in microfibril orientation from one layer to the next. The secondary wall is also incrusted with hemi-cellulose, and deposits of lignin and other substances. The S3 layer may be overlaid inside the cell lumen by a covering termed the *warty layer* (figs. 5 and 6). This layer when present, so named on account of its small protrusions, is laid down just prior to death of the cell protoplast and covers the entire S3 layer, pit cavities and any other wall scupturings.

Individual cells are joined together by intercellular material between their primary walls. This *middle lamella* is an amorphous mass, rich in lignin and low in cellulose, composed largely of pectin compounds. It is readily dissolved away by macerating solutions.

Further Reading

LIESE, W. 1965. 'The warty layer.' In *Cellular Ultrastructure of Woody Plants,* ed W. A. Côté, 251-269. Syracuse University Press. New York.

MARK, R. E. 1967. *Cell Wall Mechanics of Tracheids.* Yale University Press, New Haven.

PRESTON, R. D. 1952. *The Molecular Architecture of Plant Cell Walls.* Chapman and Hall, London.

WARDROP, A. B. 1964 'The structure and formation of the cell wall in xylem.' In *The Formation of Wood in Forest Trees,* ed M. H. Zimmerman, 87–134. Academic Press, New York.

Figure 3. *A schematic diagram to illustrate the structure of the plant woody cell wall and the orientation of the cellulose microfibrils within the various layers. ML, middle lamella; P, primary wall; S1, outer layer of the secondary wall; S2, middle layer of the secondary wall; S3, innermost layer of the secondary wall; HT, helical thickening; W, warty layer.*

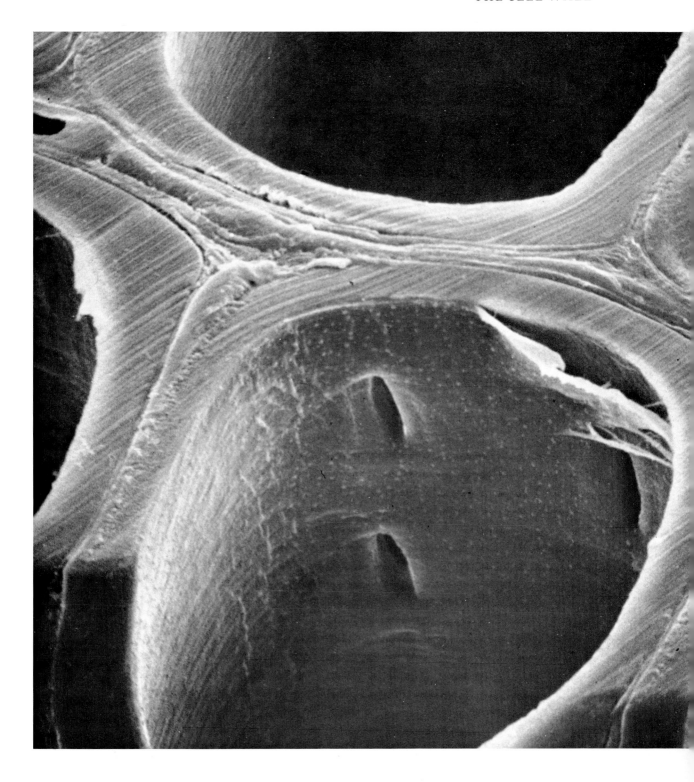

Figure 4. *The wall structure in tracheids of* Pinus radiata *D. Don. The middle lamella and primary walls can be seen to be overlaid by a thick secondary wall. In this particular photograph the S3 layer appears to have been separated from the S2 layer when the transverse surface cut was made. The warty layer can just be distinguished on the surface of the secondary wall.* (× 4500)

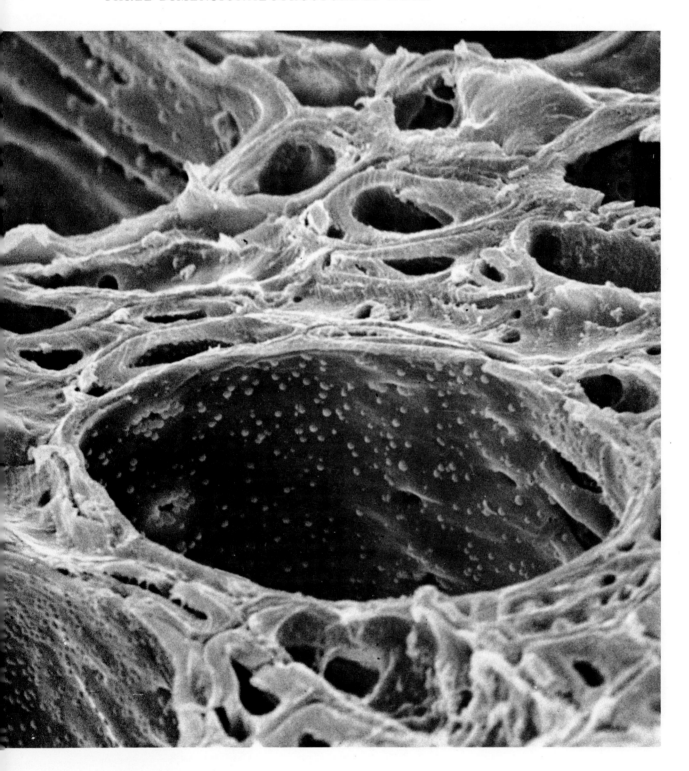

Figure 5. *The secondary walls of vessels in beech* (Fagus sylvatica) *overlaid by a warty layer.* (× 3600)

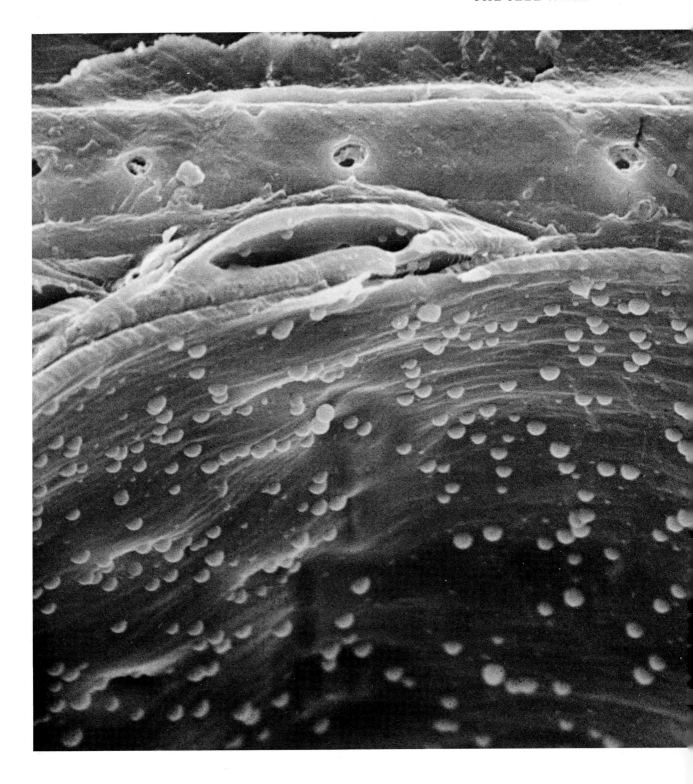

Figure 6. *A close up surface view of the warty layer on a vessel member wall in beech* (Fagus sylvatica). (× 6200)

Pits

Intervascular pit-pairs are a characteristic feature of the walls of xylem cells. These are more conspicuous in wood cells than in other plant cells because of the thick secondary walls present in the xylem. Pits are areas through which sap may pass from one cell to another and are normally found in complementary pairs, one in each of the adjacent cells forming a *pit-pair*. Each half of the pit-pair is separated by the *pit membrane* formed from the middle lamella and primary walls of the two cells (figs. 8 and 9). The presence of this membrane distinguishes a pit from a *perforation* where there is a complete opening from one cell to another. The term *pit* is often used loosely to describe the entire structure including the depression in the cell wall, the pit membrane and that portion of the secondary wall that surrounds the opening.

Two principal types of pit occur, these are *bordered pits* and *simple pits*. In the first type, a border of secondary wall material overarches the pit membrane. This has the effect of producing a maximum area of pit membrane with a minimum loss of rigidity to the cell wall. In the simple pit there is no such overarching by the secondary wall and the opening remains more or less the same diameter for the full depth of the pit. When the complementary pits of two adjacent cells are both simple they form a *simple pit-pair*, when they are both bordered they form a *bordered pit-pair*, and when one is simple and one is bordered they form a *half-bordered pit-pair*. When a pit has no complementary pit in the neighbouring cell or opens into an intercellular space it is termed a *blind pit*. *Unilateral compound pitting* results when two small pits in one cell are paired with one larger pit in the adjacent cell (fig. 10).

(a) *Bordered pits*

Bordered pits are found on the walls of hardwood vessels, tracheids and fibres and in a more specialized form on the walls of conifer tracheids. They are formed by an overarching of the pit membrane by the secondary wall. The cavity between the pit membrane and the overarching border is termed the *pit chamber* or the *pit cavity* and the opening in the actual secondary wall is termed the *pit aperture*. When the wall is very thick, as in some late wood tracheids, the opening between the cell lumen and the pit chamber may become extended into the *pit canal* with an *inner aperture* opening into the cell lumen and an *outer aperture* opening into the pit chamber (fig. 7).

The most specialized type of bordered pit is that found on the walls of conifer tracheids. Here the pit border is raised above the level of the rest of the secondary wall (figs. 7a and 11). The pit membrane is also a modified structure composed mostly of radially orientated microfibrils with a central thickened area, usually lens shaped, known

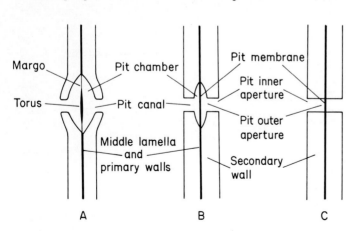

Figure 7. *The common types of pit-pair.* **A**, *conifer bordered pit where the border is raised above the level of the surrounding wall and the membrane has a torus;* **B**, *reduced bordered pit as found in many fibre tracheids;* **C**, *simple pit.*

as the *torus*. The thinner part of the pit membrane around the torus is termed the *margo* (fig. 12). During differentiation of the margo the matrix material is dissolved away and a direct pathway is established between the cells through the microfibril net. This entire structure acts very much like a valve. When the torus is in the central position of the pit chamber, sap may pass from one cell to the next by means of the margo and pit apertures. However, if the torus is deflected against either aperture of the pit-pair, it acts as a valve and the flow of sap is obstructed. Such a pit is said to be *aspirated* (fig. 13). When water is lost from the tracheids during air or kiln drying, aspiration of the pits usually occurs. Since this process is irreversible in dried timber, it makes the impregnation of dried wood in some species by preservatives very difficult.

Reduced bordered pits are found in latewood conifer tracheids and the cells, especially of the fibres, in hardwoods. These are rather simpler than the normal conifer bordered pit. The pit is still divided into a pit canal and a pit chamber (fig. 7b), but the border is not raised above the level of the rest of the secondary wall of the cell and there is no torus present on the pit membrane. The inner pit aperture, however, may show considerable variation in outline. When they are slit-like, it is quite common to find inner apertures of the pit-pair at right angles to each other forming a *cross pit-pair*.

(b) *Simple pits*

Simple pits are found in a variety of wood cells. Here the pit cavity is approximately uniform in diameter (fig. 7c and also figs 33 and 37 of later section) with only minor changes in width towards the cell lumen. Branched pits can result from the coalescence of two simple pits when a very thick secondary wall is laid down by the cell.

Like the bordered pit, the pit membrane of the simple pit is composed of the primary walls and the middle lamella of the adjacent cells of the pit-pair, but compared with the conifer bordered pit the microfibrils of the membrane are more randomly orientated and more loosely packed. There is no torus and no openings are visible in the membrane.

(c) *Vestured pits*

In some dicotyledonous woods, particularly members of the Leguminosae, small outgrowths or sculpturings are present in the pit cavities and on adjacent cell walls of the vessel members. At times they almost completely occlude the pit. These growths are *vestures* and pits containing them are termed *vestured pits* (figs. 14 and 15). Although they have the same appearance as the particles of the warty layer of the secondary wall, studies suggest that vestures develop while the contents of the cell are still living, whereas the warty layer appears to be deposited during the degeneration of the cytoplasm.

Vestures can be located at the entrance to the pit chamber, in the pit canal, or passing through the inner pit aperture into the cell lumen. Their function is obscure but undoubtedly they restrict the flow of sap through the pit.

Further Reading

LIESE, W. 1965. 'The fine structure of bordered pits in softwoods.' In *Cellular Ultrastructure of Woody Plants*, ed W. A. Côté, 271–290. Syracuse University Press, New York.

SCHMID, R. 1965. 'The fine structure of pits in hardwoods.' In *Cellular Ultrastructure of Woody Plants*, ed W. A. Côté, 291–304. Syracuse University Press, New York.

SCHMID, R. and MACHADO, R. D. 1968. 'Pit membranes in hardwoods – fine structure and development.' *Protoplasma*, **66**, 185-204.

SCURFIELD, G. and SILVA, S. R. 1970. 'The vestured pits of *Eucalyptus regnans* F. Muell: a study using scanning electron microscopy.' *Botanical Journal of the Linnean Society*, **63**, 313-320.

SCURFIELD, G. and SILVA, S. R. 1970. 'Vessel wall structure: an investigation using scanning electron microscopy.' *Australian Journal of Botany*, **18**, 301-312.

TSOUMIS, G. 1965. 'Light and electron microscopic evidence on the structure of the membrane of bordered pits in tracheids of conifers.' *In Cellular Ultrastructure of Woody Plants*, ed W. A. Côté, 305–317. Syracuse University Press, New York.

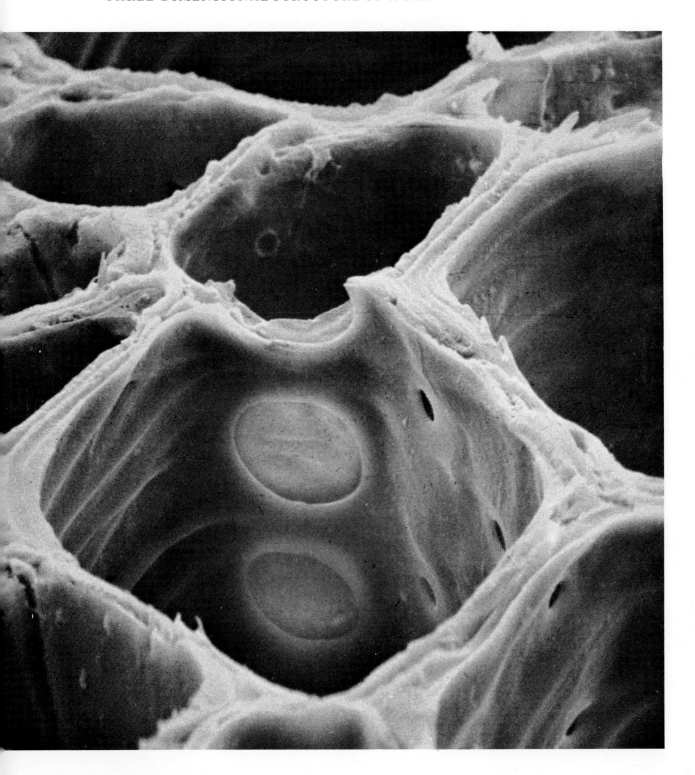

Figure 8. *Pit-pairs in adjoining vessels in* Pseudopanax arboreum *(Murr)*. *In this photograph the thin pit membrane is intact in the lower pit-pairs but in the top one it has been torn while making the transverse cut.* (× 4700)

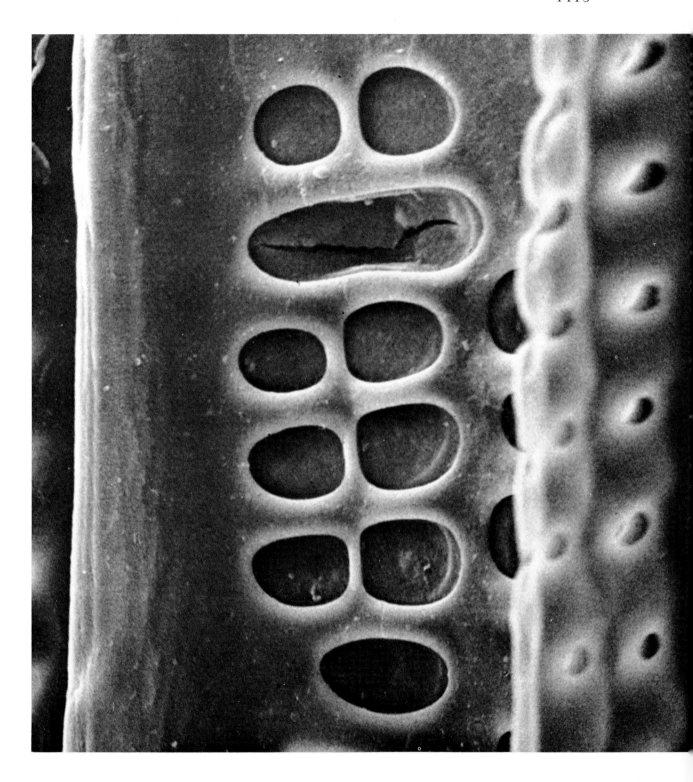

Figure 9. *Pits in the secondary wall viewed from inside a vessel element of red beech* (Nothofagus fusca *Hook f.*). *The pit membranes are intact in the circular pits, but the membrane has ruptured in the oblong-shaped pit.* (× 3400)

THREE-DIMENSIONAL STRUCTURE OF WOOD

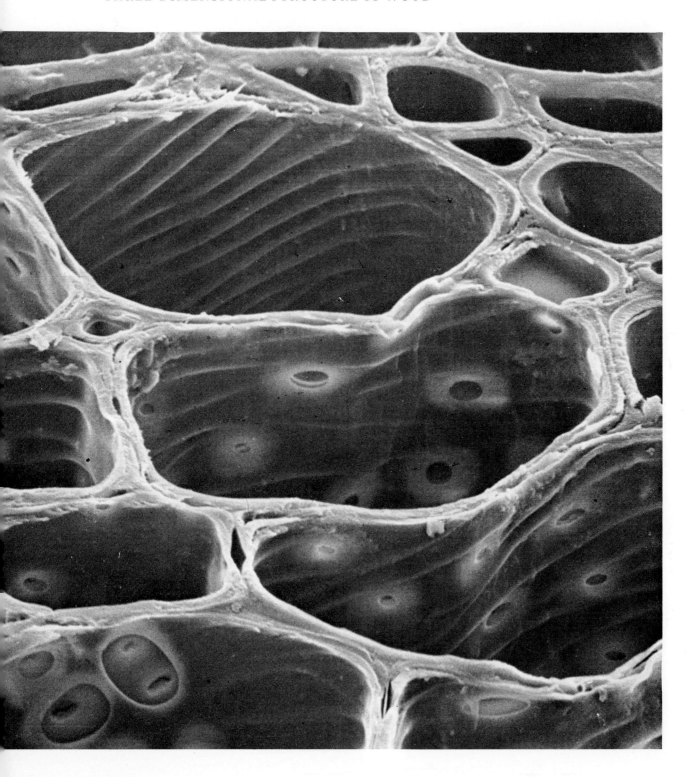

Figure 10. *In this transverse section of* Pseudopanax arboreum *circular pits can be seen on the vessel walls. In the lower left, two pits on the farther cell can be seen to complement the single larger pit in the nearer cell. This is an example of unilateral compound pitting. Helical thickening is visible on the walls of the vessel members of this wood.* (× 1350)

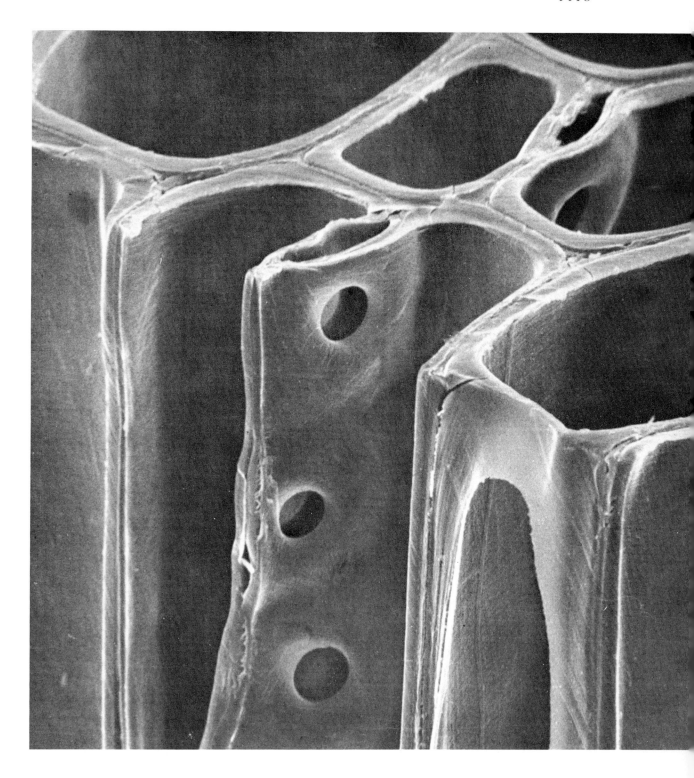

Figure 11. Bordered pits seen on the radial walls of these tracheids in Pinus radiata. The pit apertures are clearly visible and the transverse section has cut through two pit chambers revealing the overarching secondary walls. (× 2100)

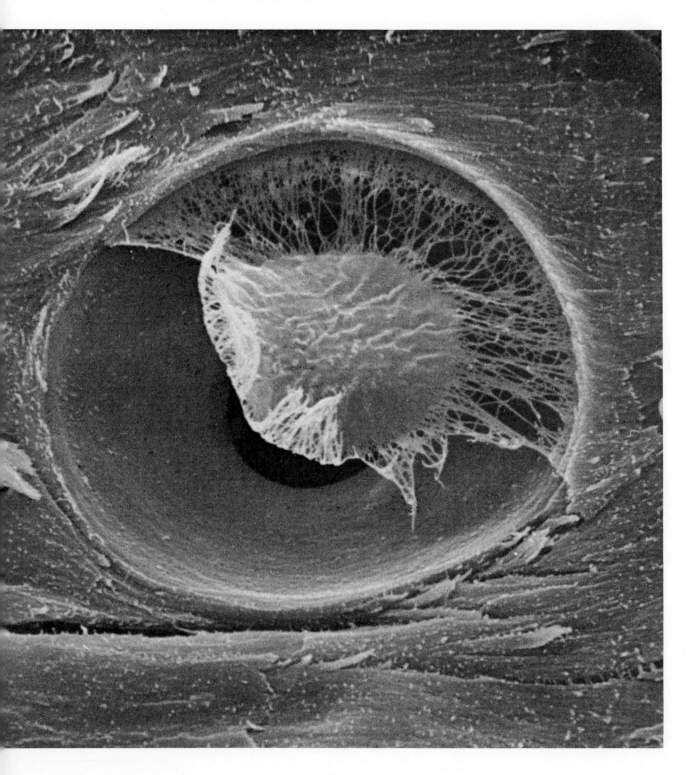

Figure 12. *A radial longitudinal cut in the primary wall area has exposed this torus, suspended by the radially orientated microfibrils of the margo. These microfibrils have separated from the tracheid wall in the lower half of the photograph when the cut was made. The torus appears to be unevenly thickened.* (× 8200)

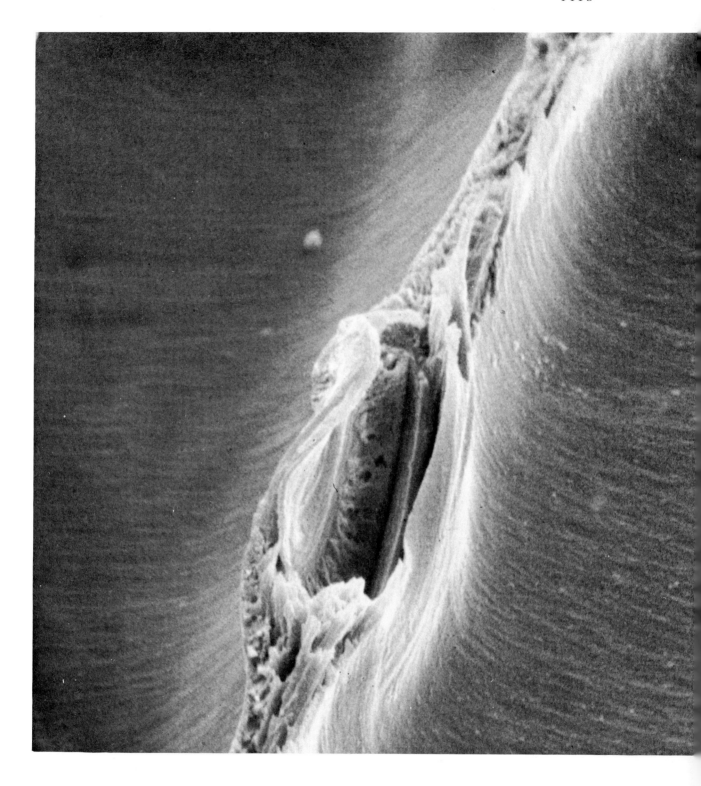

Figure 13. *A section through an aspirated pit in* Pinus. *In this view the pit chamber can be seen with the torus deflected against the pit aperture of the cell to the left, thus obstructing the flow of sap from the cell on the right.* (× 6100)

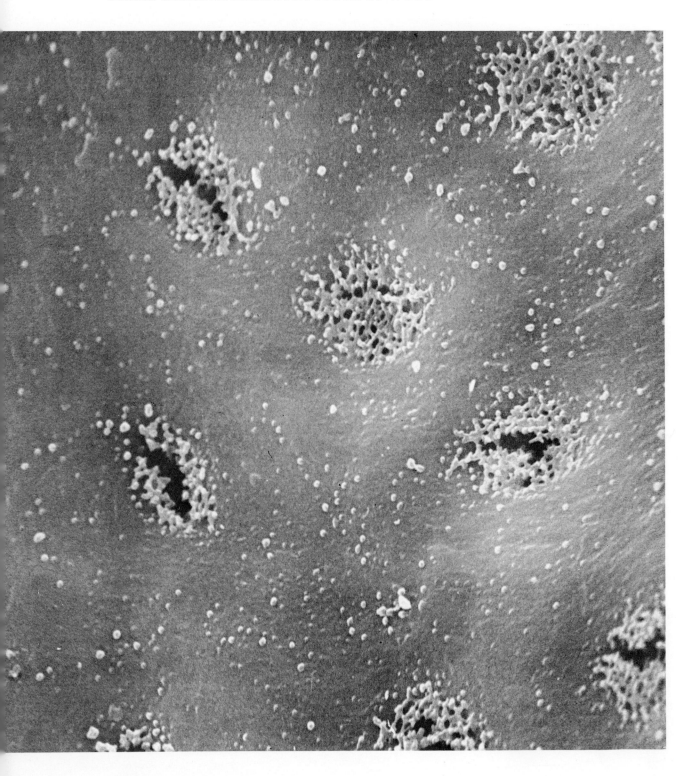

Figure 14. *Vestured pits seen in surface view in Eucalyptus delagatensis R.T.B. Note the similarity between the vestures and the particles of the warty layer when seen in this view.* (× 7500)

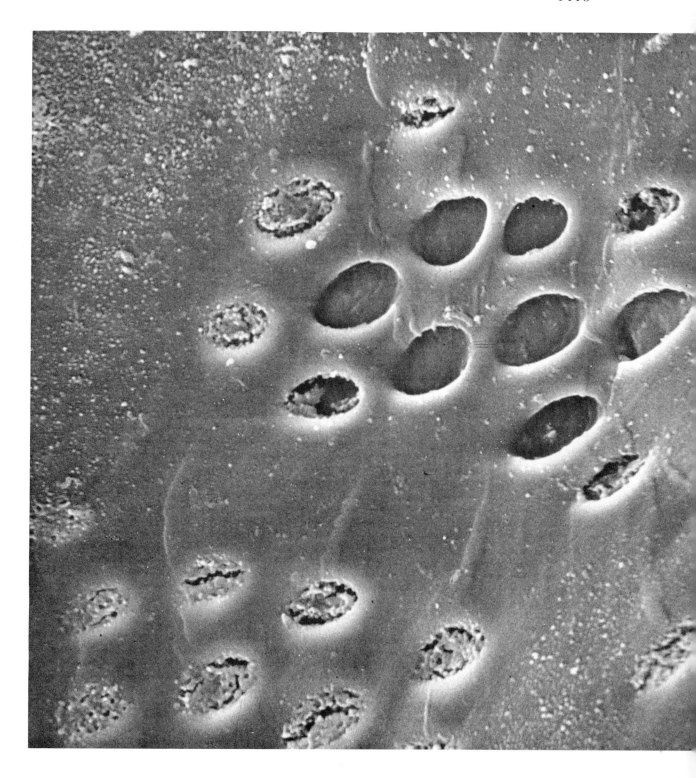

Figure 15. *Vestured, partly vestured and non-vestured pits in a vessel member of northern rata* (Metrosideros robusta *A. Cunn.*). *The warty layer is more dense in some areas.* (× 4100)

Distribution and patterns of wall pitting

There can be considerable variation in the size and distribution of the pits on the walls of wood cells. They may occur on any part of the cell wall or they may be more concentrated towards the tips of the cell. In conifer tracheids (fig. 11), the bordered pits occur more commonly on the radial walls, but in hardwood cells the pits may cover any wall of the cell. The inner pit apertures can also illustrate various outlines, being circular, elliptical or forming slits in the wall. In *scalariform* pitting the apertures are all elliptical and the pits are arranged in rows with their long axes in the transverse plane of the cell (fig. 16). When the pits are more rounded in outline the pitting may be *irregular* if the inner apertures show no pattern, or *opposite* if they are arranged in regular rows (fig. 18). In a few plants the pits are so crowded that their borders form hexagonal outlines. This *alternate* distribution of pitting is found in *Agathis* and *Araucaria* (fig. 17).

Softwood tracheids generally have the more specialized conifer bordered pitting between adjacent tracheids, but tracheid to parenchyma pitting is of the half-bordered type with the simple pit on the parenchyma cell side. Although generally simpler in structure, the pit-pairs of the hardwoods are more varied than the bordered pits found in softwoods, the larger number of cell types present in hardwoods resulting in a greater diversity of pit-pair combinations (figs. 18 to 20). Vessel members may possess simple or bordered pits without tori, hardwood tracheids and fibre tracheids usually have reduced bordered pits whereas in libriform fibres the pitting is generally simple.

Further Reading

ESAU, K. 1965 *Plant Anatomy*. Wiley, New York.
FAHN, A. 1967. *Plant Anatomy*. Pergamon Press, Oxford.

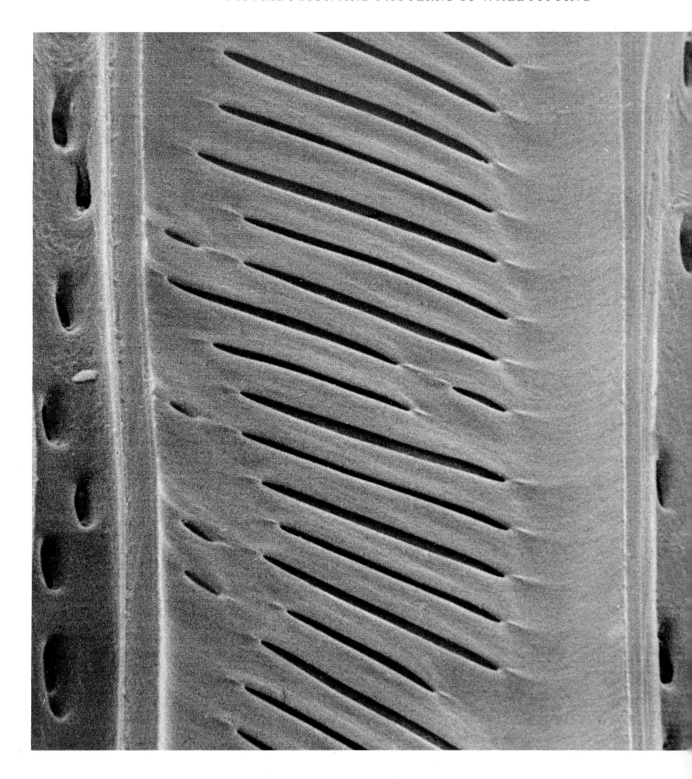

Figure 16. *Scalariform pits in a vessel element of* Laurelia novae-zelandiae. (× 1950)

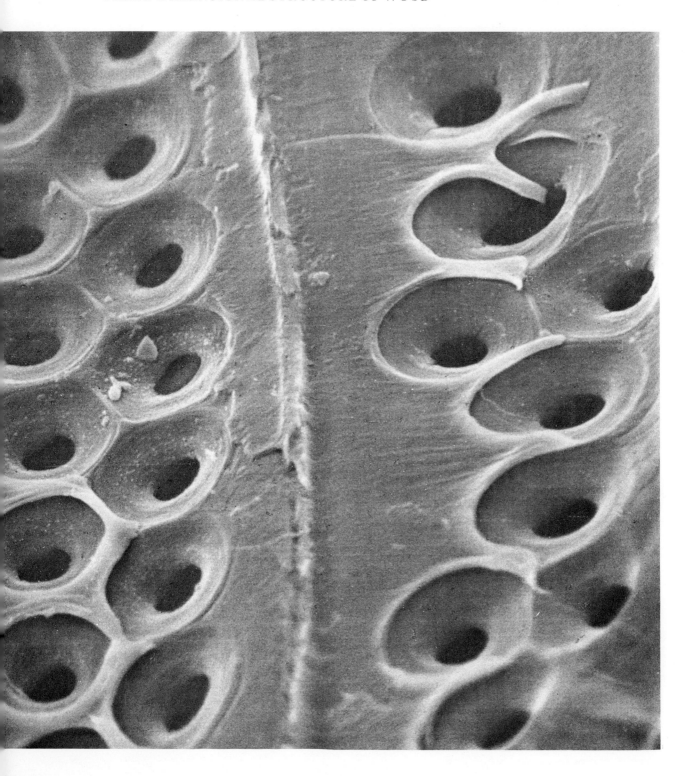

Figure 17. *Alternate pitting on the tracheid walls of kauri* (Agathis australis *Salisb.*). (× 3300)

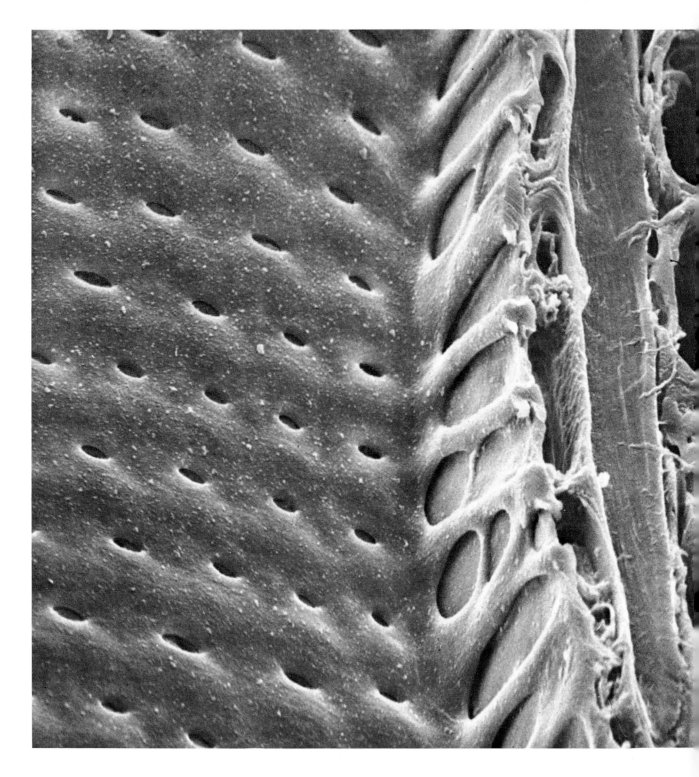

Figure 18. *Vessel to vessel and vessel to ray pitting in red beech* (Nothofagus fusca). *The neat rows of reduced bordered pits on the left connect with complementary pits in an adjacent vessel element, while the larger ones on the right hand wall connect with the pits in the neighbouring ray cells.* (× 2600)

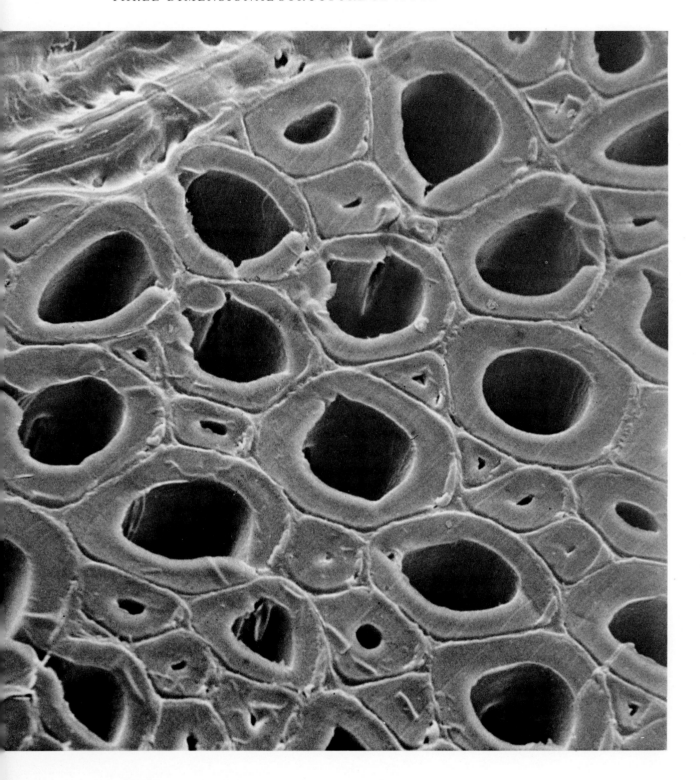

Figure 19. *Reduced bordered pit-pairs connecting fibres in* Beilschmedia tawa *(A. Cunn.). The inner apertures are slit shaped.* (× 2300)

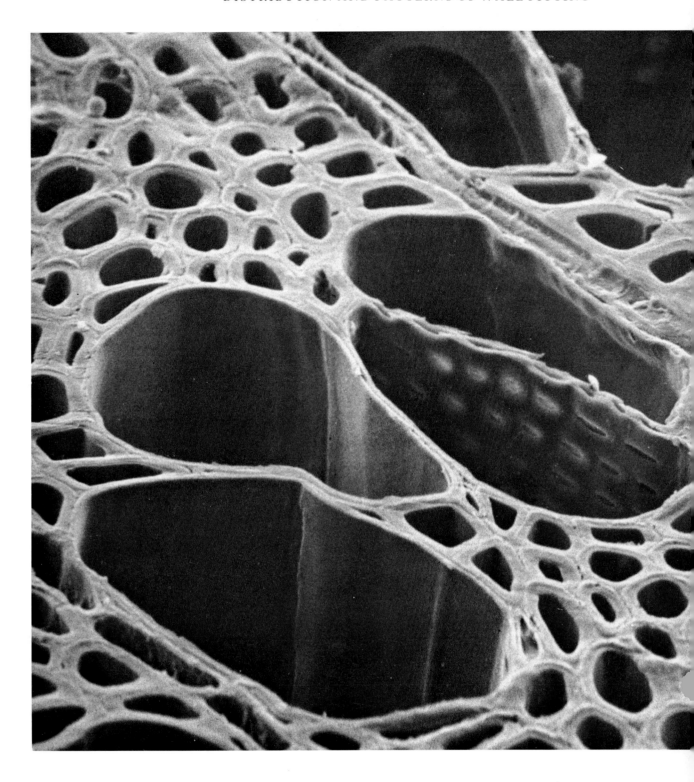

Figure 20. Vessel member walls are not always covered in pits. In this view of a cluster of vessels in red beech, some walls show prolific vessel to vessel pitting while others are completely devoid of pits. (× 1250)

Other wall sculpturing

In addition to the pits, the walls of some tracheary cells may be further modified by the presence of various forms of thickening occurring on the inside of the secondary wall. Bands of *helical thickening* are present in the tracheids of a few conifers, notably Douglas fir (figs. 21 and 22). They may occur also in the ray tracheids in the late wood of some pines. The angle of the helix appears to be related to the size of the cell, the longer and thinner cells having the more open helices. Helical thickening also occurs on the inside of the secondary wall in a number of hardwood vessels (figs 23 and 24). These thickenings are considered to be extensions of the S3 layer of the secondary wall.

Further Reading

CÔTÉ, W. A. 1967. *Wood Ultrastructure*. University of Washington Press, Seattle.

WARDROP, A. B. and DADSWELL, H. E. 1951. 'Helical thickenings and micellar orientation in the secondary wall of conifer tracheids.' *Nature*, **168**, 610-612.

OTHER WALL SCULPTURING

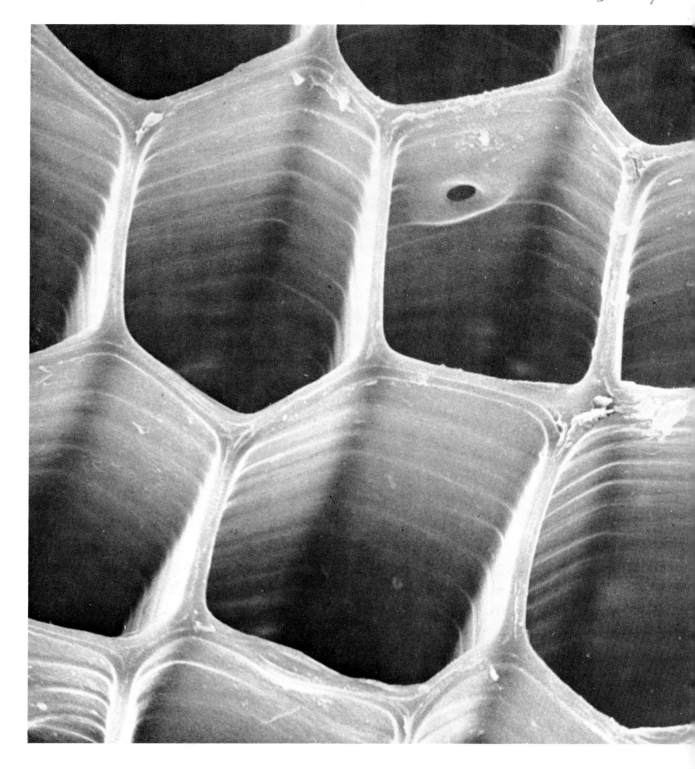

Figure 21. *A view looking down into the tracheids of Douglas fir* (Pseudotsuga menziesii *Franco), showing the helical thickening on the secondary walls. Note the single conifer bordered pit.* (× 1750)

Figure 22. *A closer view of the helical thickenings on the secondary walls of a Douglas fir* (Pseudotsuga menziesii) *tracheid.* (× 10,000)

OTHER WALL SCULPTURING

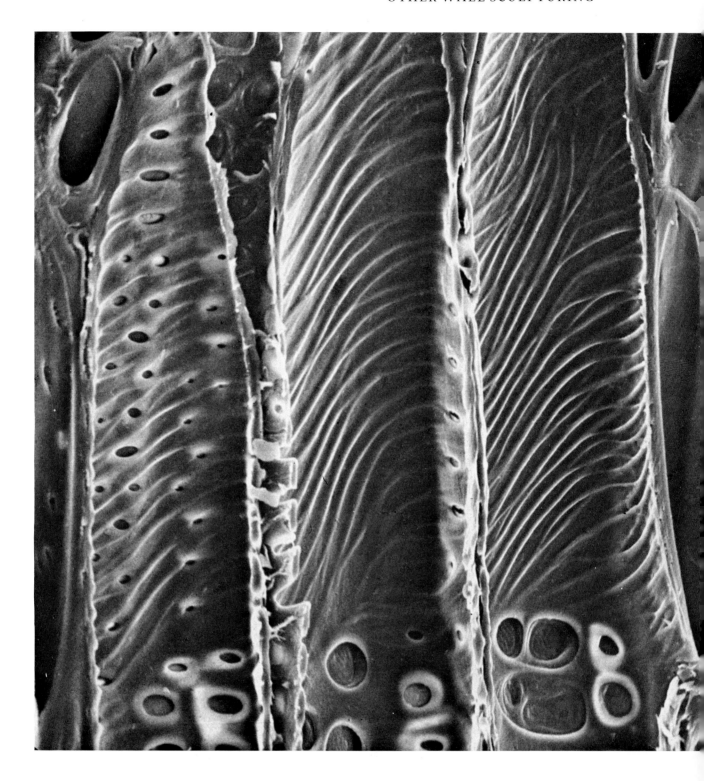

Figure 23. *Helical thickenings on the vessel walls of* Pseudopanax arboreum. *The concentration of larger pits at the bottom of the photograph (cross-field pitting) marks the position of a ray passing across behind the vessels.* (× 1300)

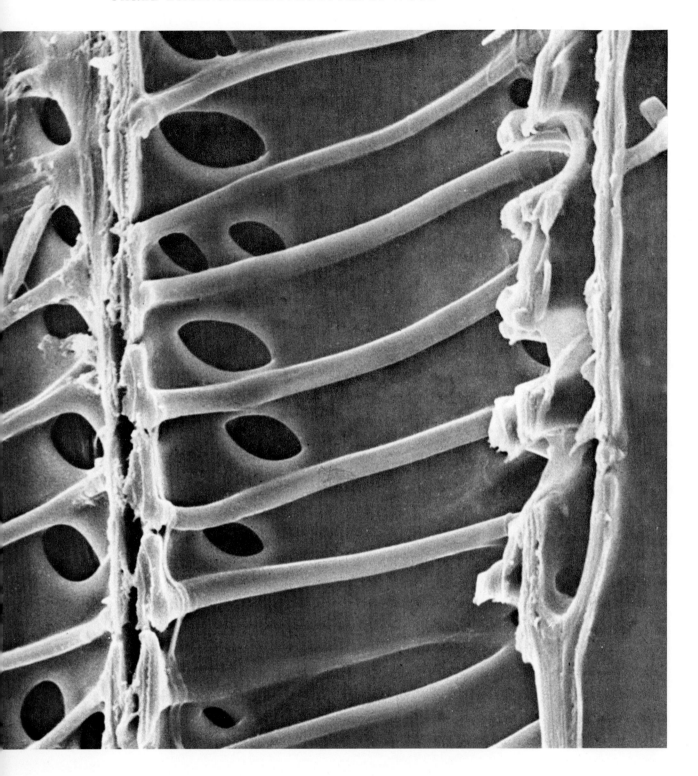

Figure 24. *Helical thickening on the vessel member walls of lacebark* (Hoheria populnea A. Cunn.). (× 3500)

Perforation plates

Vessel members are normally joined together end to end to form long continuous tubes called vessels (fig. 25). Vessels may extend quite short distances in some plants, while in others they may extend the full height of the tree. Sap moving through these structures passes freely from cell to cell through perforations in the cell walls. These perforations are relatively simple openings in the cell wall compared with the more complex structure of pits where a membrane may still be present. Although these perforations can occur over any part of the vessel wall, they are usually more frequent on the end walls of the cell, forming what is termed a *perforation plate*. A perforation plate containing a single large opening is termed a *simple perforation plate* (fig. 26). Where a number of openings are grouped together it is termed a *multiple* perforation plate. The openings in a multiple perforation plate may be arranged as a series of slits to form a *scalariform* perforation plate (fig. 27) or they may be arranged in a more irregular or net form termed a *reticulate* perforation plate (fig. 28).

Simple perforation plates are thought to have evolved from the multiple type by the loss of the bars of thickening. In the most advanced vessels, all that remains of the perforation plate is a small rim around the inside of the vessel (fig. 29). The end walls of each vessel member are generally sloping, the angle of the slope varying with the plant species.

Further Reading

ESAU, K. 1965. *Plant Anatomy*. Wiley, New York.
ESAU, K. and HEWITT, W. M. B. 1940. 'Structure of end walls in differentiating vessels.' *Hillgardia*, **13**, 229-244.
GREENIDGE, K. N. H. 1952. 'An approach to the study of vessel length in hardwood species.' *American Journal of Botany*, **39**, 570-574.

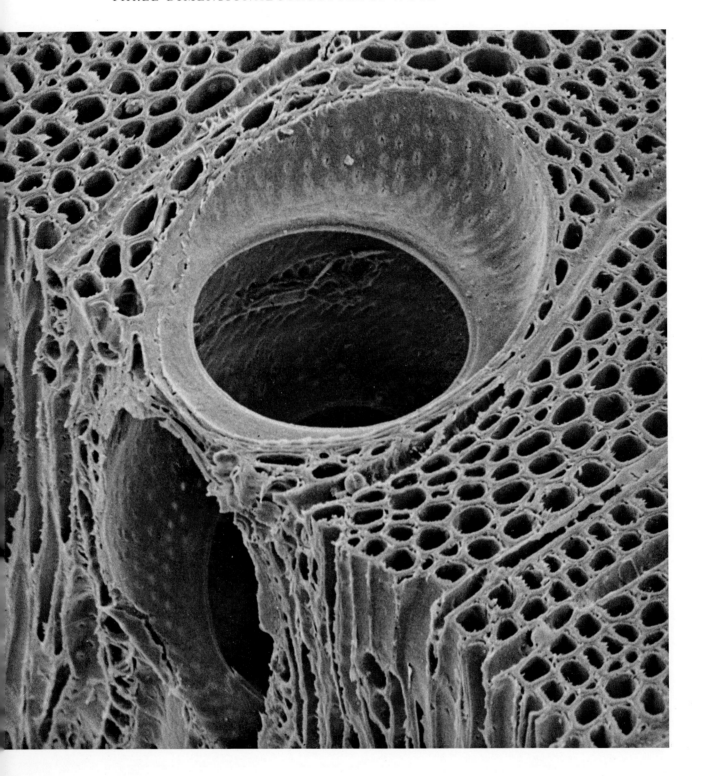

Figure 25. *Vessel members joined end to end to form a vessel can be seen in this transverse and tangential longitudinal view of* Eucalyptus delagatenis *wood. The vessel members have simple perforation plates, one can be seen in the transverse section and a second is visible, partly through the top perforation and partly through the side of the vessel member which has been cut open.* (× 600)

PERFORATION PLATES

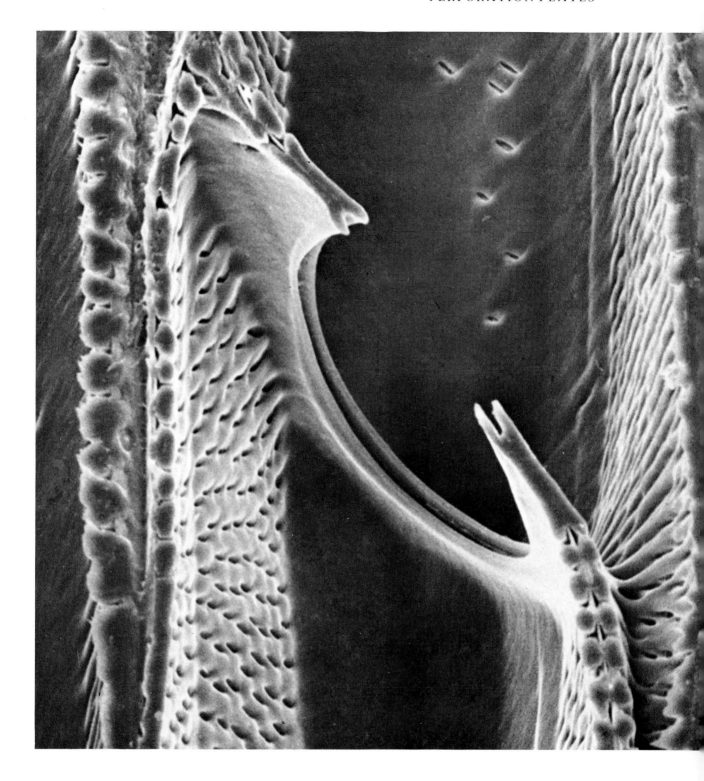

Figure 26. *A simple perforation plate between vessel members in the New Zealand honeysuckle* (Knightia excelsa *R.Br.). Notice the pit membranes traversing the pit-pairs on the side walls and the oblique angle of the perforation plate with its slight border.* (× 1700)

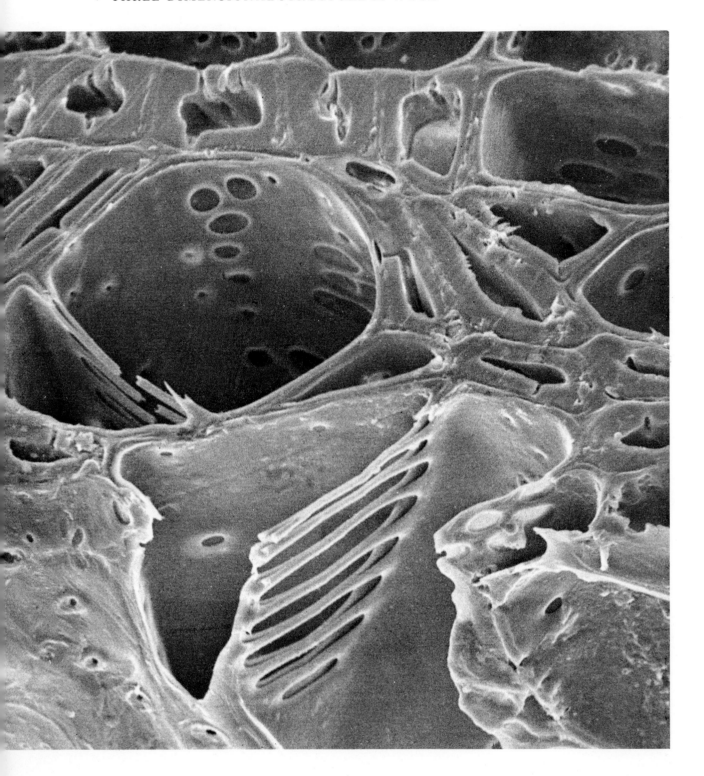

Figure 27. *A scalariform perforation plate in* Griselinia littoralis *Raoul. Very reduced bordered pits can be seen in the fibres and larger ones in the vessel members.* (× 1250)

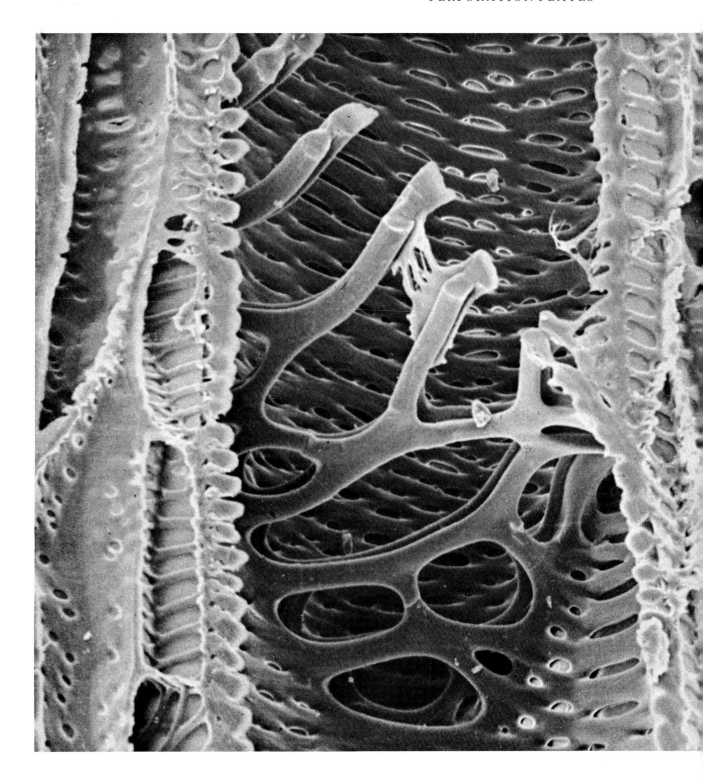

Figure 28. *A view of a reticulate perforation plate. These vessels are photographed in the Nikau palm* (Rhopalostylis sapida). *Since this plant does not have a vascular cambium this wood is not secondary xylem.* (× 900)

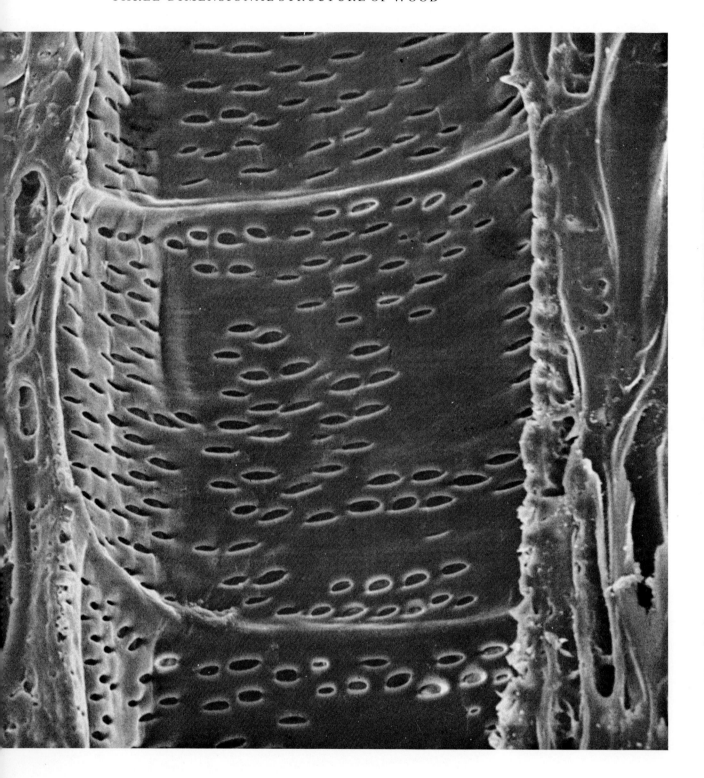

Figure 29. *Three vessel members joined end to end in a vessel of* Ulmus *wood. The simple perforation plates are so large that all that remains of them is a small rim around the vessel wall.* (× 1300)

The structure of rays

The arrangement of cells into the axial and ray systems constitutes one of the characteristic features of secondary xylem. While the axial system is built up of the tracheary elements, fibres and axial parenchyma cells, the ray system is built up largely of parenchyma cells. The cells of both systems are closely interconnected by means of pits. Since rays run radially out through the wood transverse and radial longitudinal wood sections expose their long axes, whereas a tangential longitudinal section exposes a transverse or end view of the ray.

Rays vary considerably in their size. Their height with respect to the longitudinal axis of the stem varies both between different woods and between different rays within the same wood sample. The ray height, however, is normally several times greater than the ray width. *Uniseriate* rays are only one cell in width (fig. 30), while *multiseriate* rays are more than one cell wide (figs. 31 and 32). All multiseriate rays taper towards their upper and lower margins, where they may show uniseriate extensions. Both uniseriate and multiseriate rays may occur in the one type of wood, but in others the rays are all the uniseriate variety. Some dicotyledonous woods possess two distinctly different shaped parenchyma cells in their rays. While most cells are of the radially elongated or *procumbent* type, the cells at the upper and lower margins of the rays may be elongated in the axis of the stem and are referred to as *upright* ray cells. Rays with only procumbent parenchyma cells are classified as *homocellular* rays, while those with both types of cell are grouped as *heterocellular* rays.

Ray parenchyma cells may or may not have secondary walls. When a secondary wall is present it is normally dotted with simple pits (figs. 33 and 34). The pairing of these ray pits with those on the walls of adjacent tracheary elements can result in local groupings of the pits on the walls of the tracheary cells. These *cross fields* are particularly noticeable in wide vessel members (fig. 35).

Further Reading

JANE, F. W. 1970. *The Structure of Wood*. Black, London.

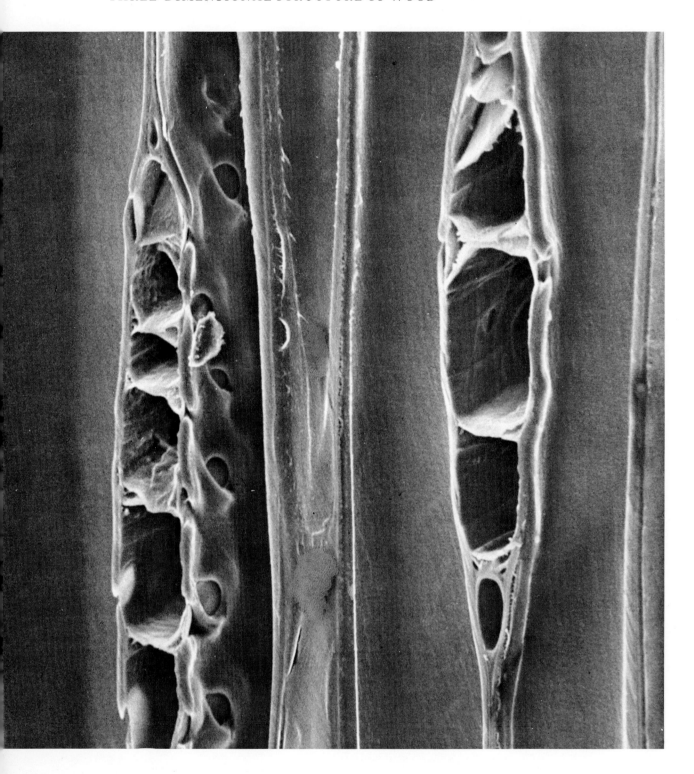

Figure 30. *Uniseriate rays in* Pinus radiata *in tangential longitudinal section. The thin-walled ray parenchyma cells show some distortion in this photograph due to the drying out of the wood specimen.* (× 1600)

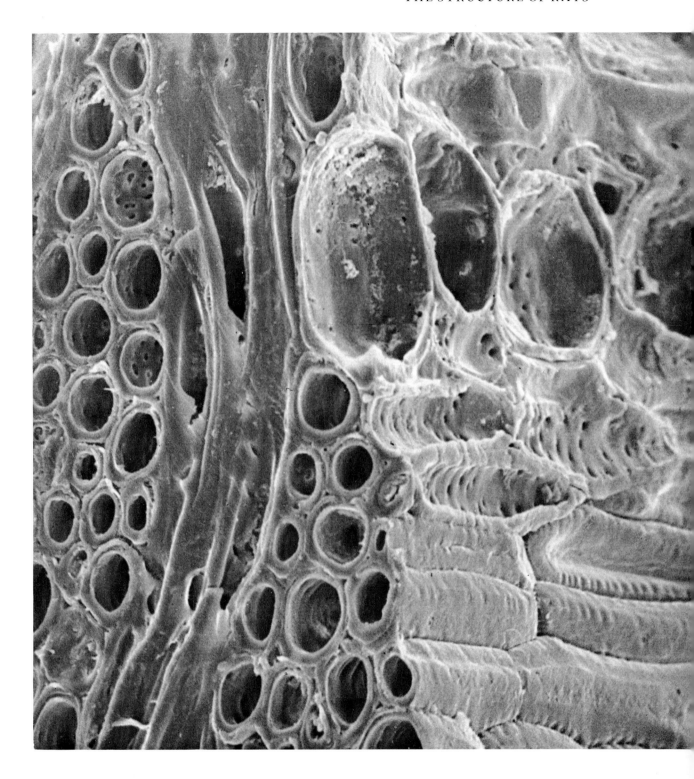

Figure 31. *Multiseriate rays in beech* (Fagus sylvatica). *Parts of two rays are seen in the tangential plane and in the radial plane the procumbent parenchyma cells can be seen arranged with their long axes in the horizontal direction.* (× 1600)

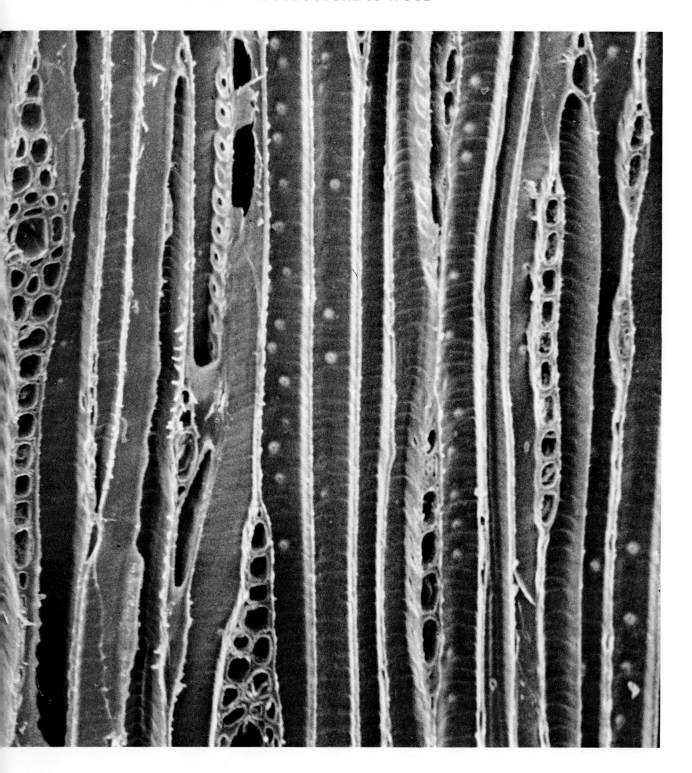

Figure 32. *Multiseriate rays with horizontal resin canals and uniseriate rays in tangential section of Douglas fir* (Pseudotsuga menziesii). *Conifer bordered pitting and helical thickening can be seen on the tracheid walls.* (× 500)

Figure 33. Procumbent ray parenchyma cells in radial longitudinal section of Hoheria populnea. Secondary walls are present with numerous simple pits. (× 600)

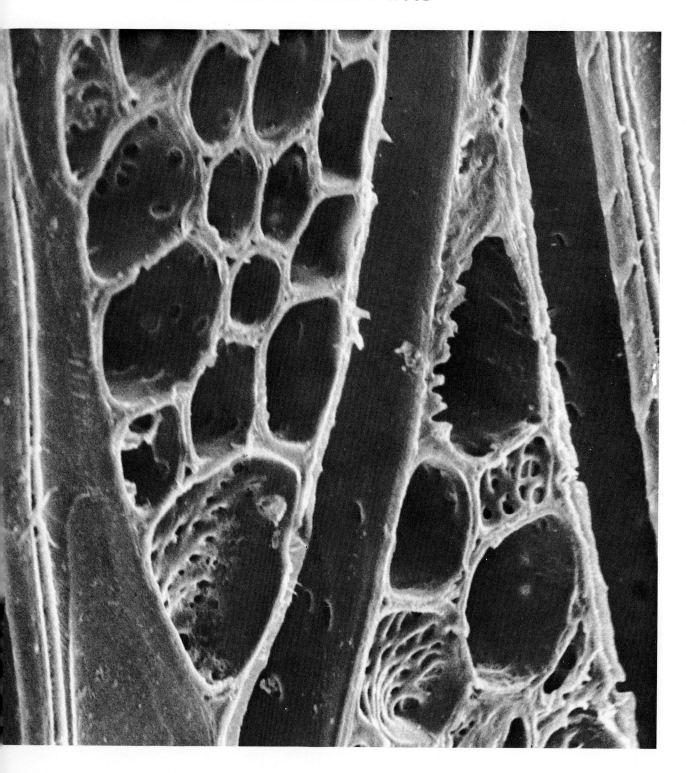

Figure 34. *A close up view of parts of two rays cut by a tangential longitudinal section of* Elaeocarpus dentatus Forst. *The end walls of the ray cells are profusely pitted.* (× 1550)

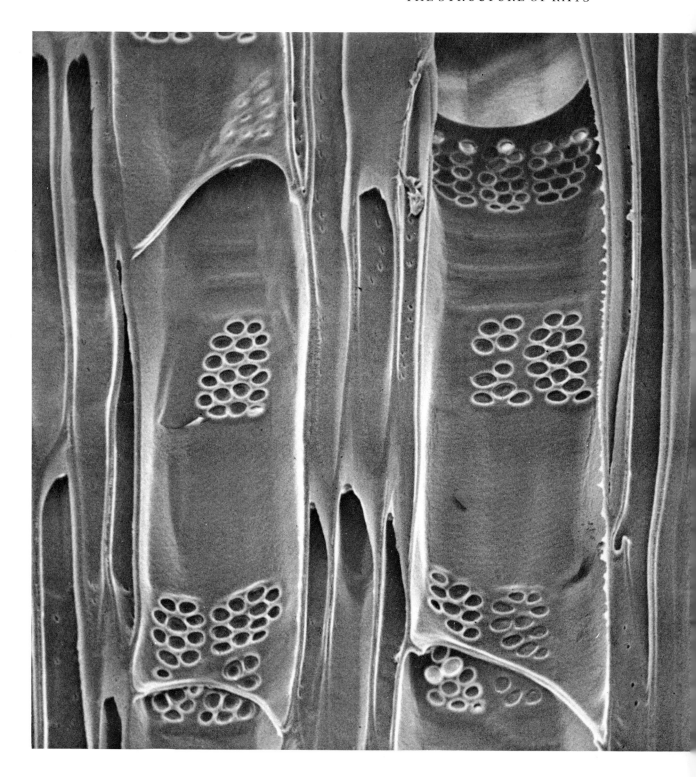

Figure 35. *Cross-field pitting seen in a radial longitudinal view of vessel members of* Populus nigra *Muench. The groups of pits on the walls of the vessel members complement the pits in the ray cells passing behind.* (× 650)

Axial parenchyma

Living parenchyma cells occur in the axial system of secondary xylem as well as in the rays. *Axial parenchyma* occurs in a number of dicotyledonous woods but is very much less frequent in gymnospermous plants. When viewed in longitudinal section its cells may occur in isolation as longish cells with pointed ends, or they may occur in *parenchyma strands* as columns of cells with almost transverse end walls (fig. 36). In transverse section they show a number of distribution patterns. When parenchyma is associated with a vessel it is described as *paratracheal* and when it is not necessarily grouped around a vessel it is *apotracheal*. A number of subdivisions of these two main patterns are often used by wood anatomists to describe the proportion and distribution pattern of axial parenchyma within the growth ring.

Like the cells of ray parenchyma, the cells of the axial parenchyma may have secondary walls (fig. 37). When present these walls are pierced by simple pits. Axial parenchyma cells in sapwood accumulate starch and other food reserves towards the end of the growing season and these are apparently utilized at the onset of cambial activity the following spring. Tannins and other substances may also occur.

Further Reading

International Association of Wood Anatomists, 1964. *Multilingual Glossary of Terms used in Wood Anatomy*.

JANE, F. W. 1970. *The Structure of Wood*. Black, London.

AXIAL PARENCHYMA

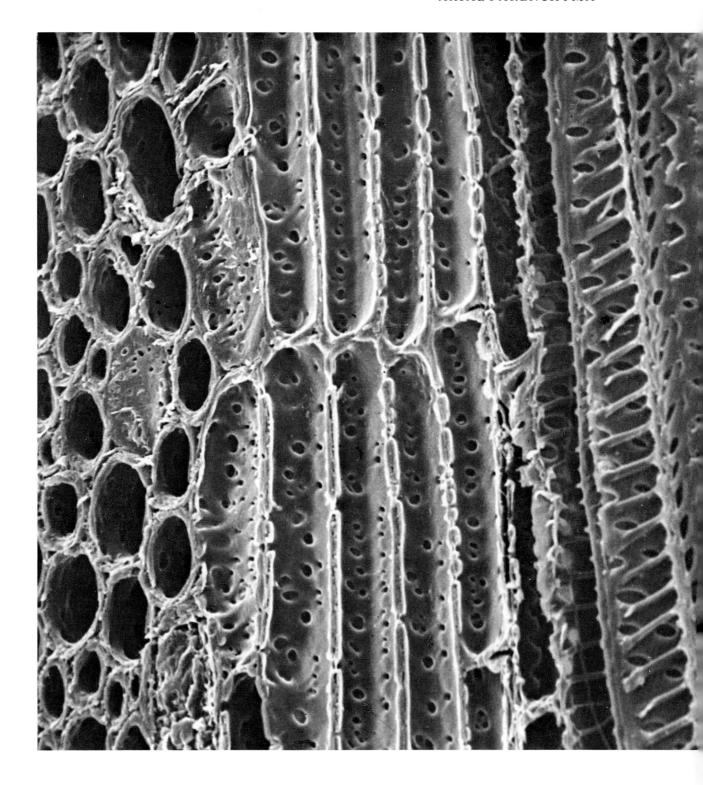

Figure 36. *Axial parenchyma cells in lacebark* (Hoheria populnea). *These cells have secondary walls with simple pits. The long cell with the helical thickening on the right of the photograph is part of a vessel member.* (× 1100)

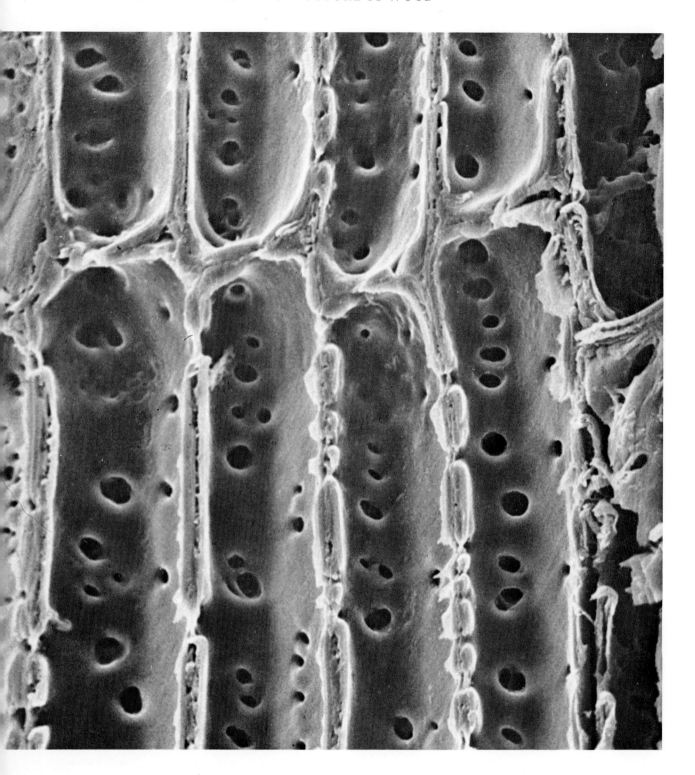

Figure 37. *A closer view of axial parenchyma cells in* Hoheria populnea. (× 2200)

Growth rings

In most climates the vascular cambium is active during the spring and summer months, but passes through a period of partial or total dormancy during the colder winter months. This cyclic pattern of activity leaves a series of rings in the secondary xylem. In most cases these *growth rings* are produced once a year, but other factors can cause the cambium to cease its activity at inbetween times of the year so that not all growth rings are necessarily *annual rings*. The cells formed during the early period of active growth constitute the *earlywood* or *springwood,* and those cells cut off by the cambium towards the end of the actively growing season form the *latewood* or *summerwood*. Latewood is denser and harder than earlywood, its cells being generally greater in length, smaller in the radial dimension and possessing relatively thicker walls (fig. 38). Earlywood cells on the other hand are thin-walled and larger in diameter. Vessels, parenchyma and fibres may show characteristic patterns of distribution within individual growth rings.

Further Reading

BANNAN, M. W. 1964. 'The vascular cambium and tree-ring development.' In *Tree Growth*, ed T. T. Kozlowski, Ronald Press, New York.

PHILIPSON, W. R., WARD, J. M. and BUTTERFIELD, B. G. 1971. *The Vascular Cambium*. Chapman and Hall, London.

THREE-DIMENSIONAL STRUCTURE OF WOOD

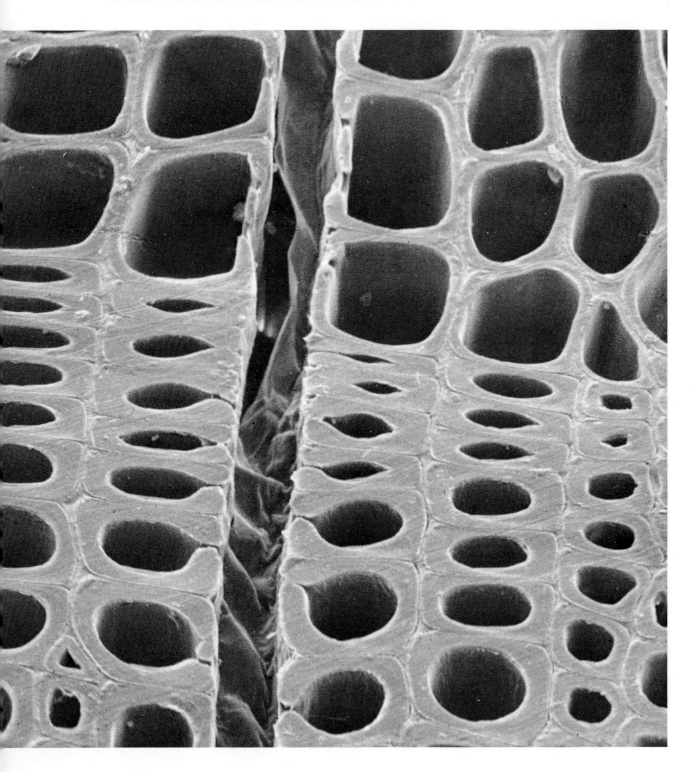

Figure 38. *An annual growth ring is clearly illustrated in this transverse section of Pinus radiata.* To the lower side small, thick walled tracheids of the latewood can be seen. Beyond the ring boundary the cells are larger and have thinner walls. The cells remain in more or less radial files after being cut off from the vascular cambium. A ray can be seen traversing the wood at right angles to the ring boundary. (× 550)

Vessel distribution

When viewed in transverse section a vessel is sometimes referred to as a *pore*. The distribution of the pores within the growth ring is a characteristic feature of woods. In many woods the vessels are more or less uniform in size and are distributed randomly throughout the growth layer. Such woods illustrate a *diffuse porous* vessel distribution (fig. 39). In others the pores are distinctly larger in the early wood than in the late wood, or may even be confined to the early wood. Such a condition is described as being *ring porous* (fig. 40).

Pores may also occur in various groupings in the wood. A pore is *solitary* when each vessel occurs singly (fig. 41). Groups of vessels are called multiples. They may be in a radial file called a *radial pore multiple* (fig. 42) or be in a more irregular arrangement termed a *pore cluster*.

A number of woods illustrate concentric bands where the derivatives of the vascular cambium have all differentiated into one type of cell. This feature invites some interesting morphogenetic enquiry as its cause is unknown. *Knightia, Hoheria* and *Plagianthus* all illustrate concentric bands of fibres alternating with bands of vessels and parenchyma cells (figs 43 and 44).

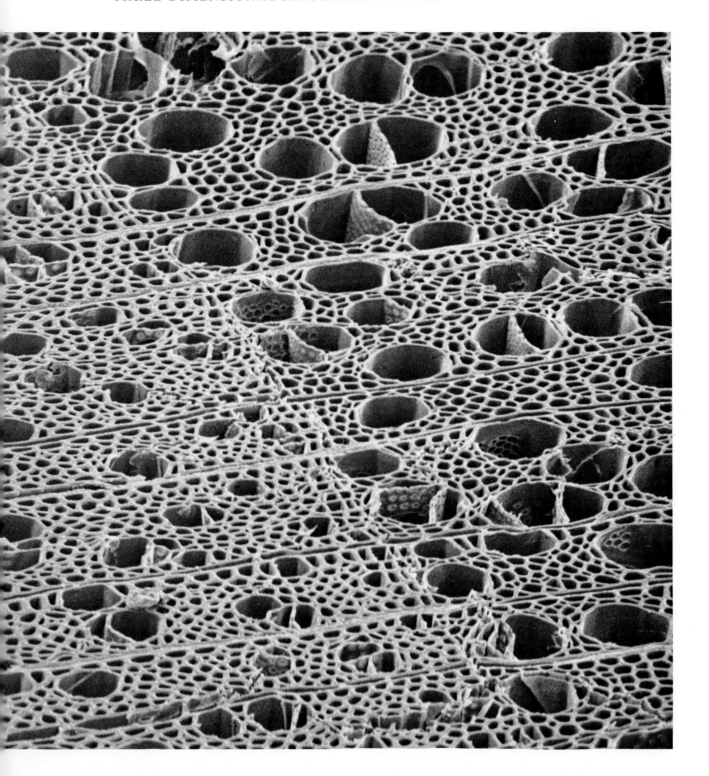

Figure 39. *A transverse section of the wood of the poplar* (Populus *sp.*). *This plant illustrates a diffuse porous distribution of vessels within the growth layer with the pores arranged singly or in multiples of twos and threes. The early wood vessels to the right of the photograph are slightly larger than those of the late wood to the left.* (× 375)

Figure 40. *Ring porous vessel distribution in the elm* (Ulmus glabra *Huds.*). *The large vessels are all grouped in the early wood.* (× 140)

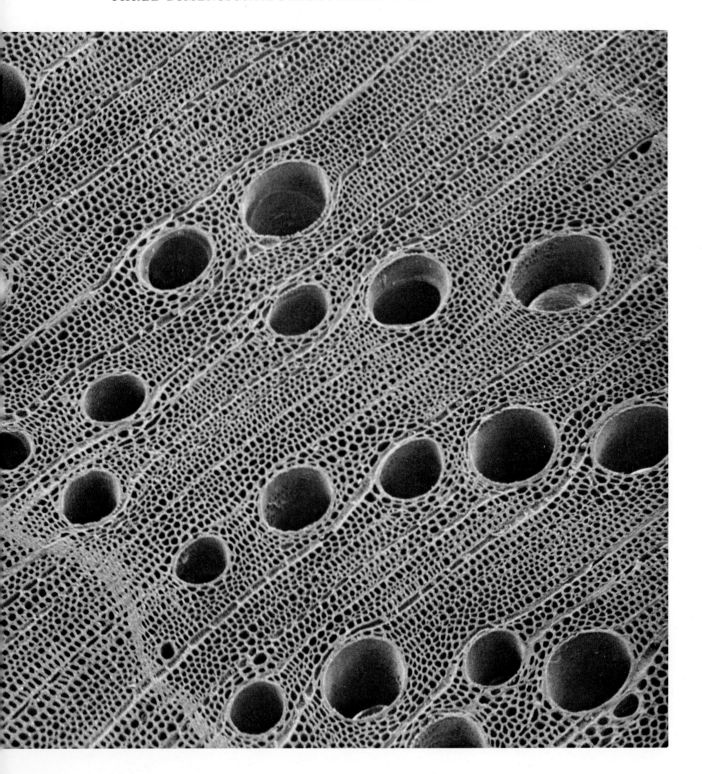

Figure 41. *Solitary vessels in* Eucalyptus delagatensis *viewed in transverse section. One complete growth ring is visible.* (× 140)

VESSEL DISTRIBUTION

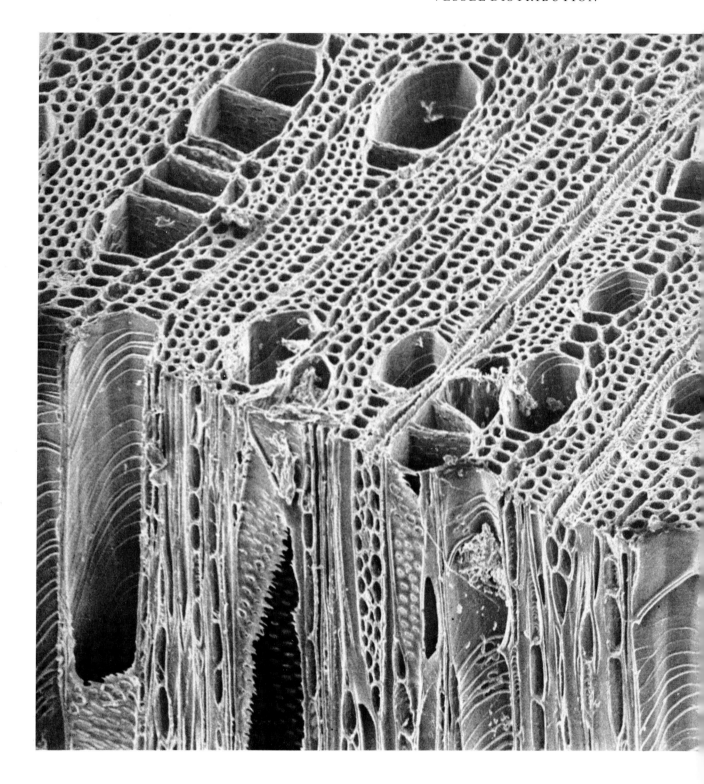

Figure 42. *Transverse and tangential views of hinau* (Elaeocarpus dentatus). *The pores in this wood are arranged in radial files. The vessel members show helical thickening on their walls and profuse vessel to vessel pitting.* (× 300)

THREE-DIMENSIONAL STRUCTURE OF WOOD

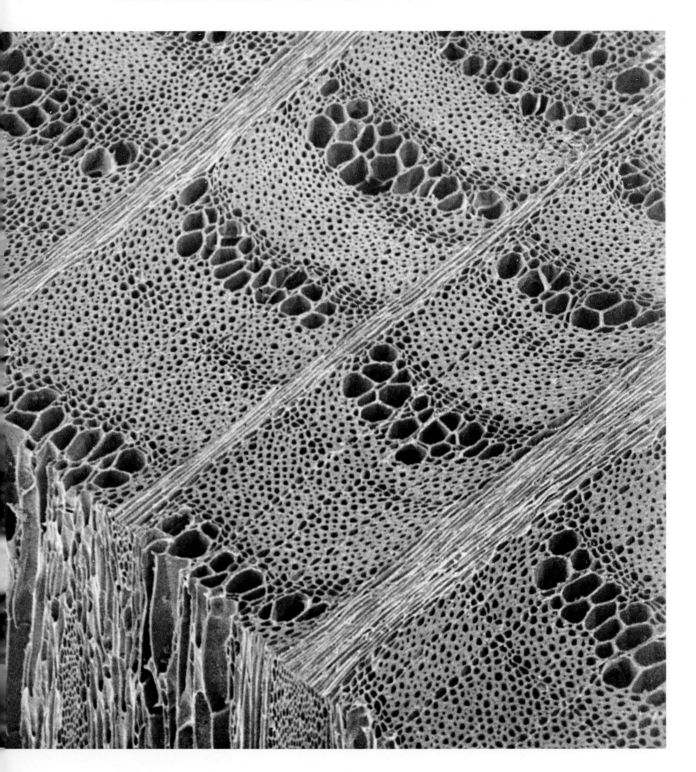

Figure 43. *Transverse view of* Knightia excelsa *illustrating the concentric bands of fibres alternating with vessels and axial parenchyma.* (× 110)

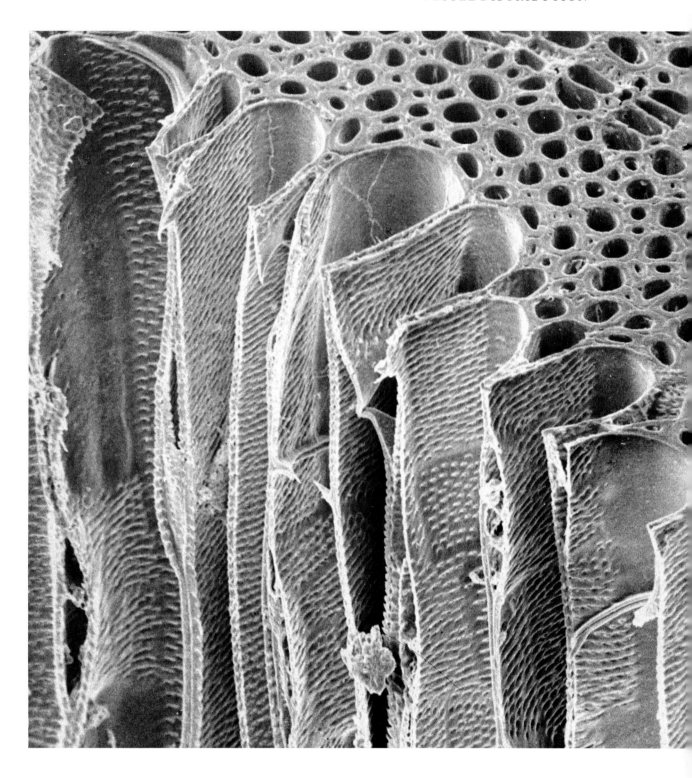

Figure 44. *A concentric band of vessels in* Knightia excelsa *seen in tangential and transverse views.* (× 500)

Resin canals

Many gymnosperm woods contain a dark coloured gummy substance called *resin*. Chemically resin is a mixture of substances, its composition varying considerably from one species to another.

Resin can be found within wood cells (intracellular) or between cells (intercellular). The former is found in both ray and axial parenchyma cells, but can also occur on the walls of tracheary elements. Intercellular resin is found in *resin canals* or *ducts*. These canals may run horizontally within rays as well as vertically among the axial xylem elements. Axial resin canals show a more or less circular outline when seen in transverse section (fig. 45), but those found in the rays may be more varied in outline (fig. 46). Resin canals are formed by the separation of axial parenchyma cells leaving an intercellular duct. The separated parenchyma cells often line the canals.

Further Reading

BANNAN, M. W. 1936. 'Vertical resin ducts in the secondary wood of the Abietineae.' *New Phytologist*, **35**, 11-46.

JANE, F. W. 1970. *The Structure of Wood*. Black, London.

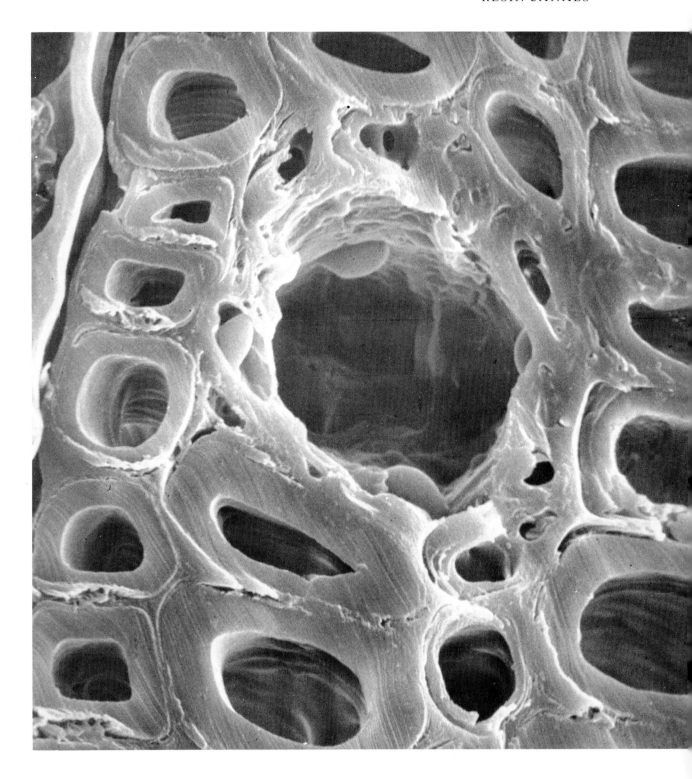

Figure 45. *An axial resin canal of Douglas fir* (Pseudotsuga menziesii) *in transverse view.* (× 2200)

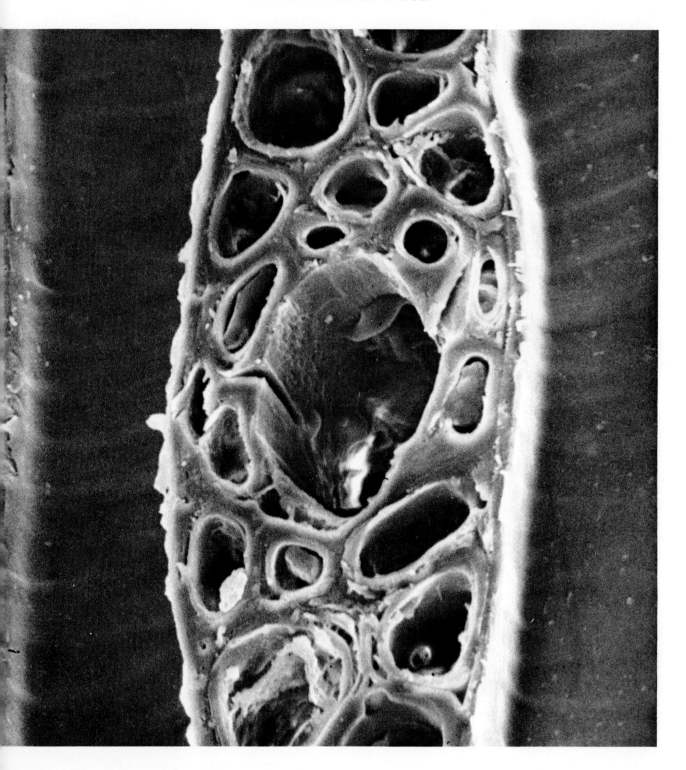

Figure 46. *A horizontal resin canal within a multiseriate ray of Douglas fir* (Pseudotsuga menziesii). (× 2500)

Tyloses

Tyloses are found as normal structures in many hardwood species. They are outgrowths of ray or axial parenchyma cells which have penetrated through the pits into the vessel lumen. For a time it was thought that tyloses developed by the out-growth of the pit membrane and parenchyma cell wall as a result of a difference in pressure with that of the empty vessel lumen. However, it now seems possible that the pit membrane is degraded enzymatically. During the growth of the proliferation, the pit membrane is lost and the parenchyma cell wall grows through the pit cavity. The cytoplasm of the parenchyma cell, including the nucleus in some cases, passes through this narrow tube into the vessel.

When this phenomenon occurs from a number of parenchyma cells through several pits into a single vessel element, the lumen of the xylem tracheary element may become completely blocked by tightly packed tyloses (figs. 47 and 48). Their shape may vary considerably, in some plants they are balloon shaped while in others they appear to be collapsed or wrinkled. Tyloses can also develop within the tracheary elements at sites of injury thus stopping the loss of sap from wounds.

After they have invaded the tracheary cells, the thin walls of the tyloses may become thickened by the deposition of a secondary wall – the process of *sclerosis*. When this happens vessel elements become filled with hard walled stone-cells.

The presence of tyloses can be a problem in wood preservation since profuse tylosis inhibits the penetration of wood by preservatives.

Further Reading

ISHIDA, S. and OHTANI, J. 1969. 'Study of tyloses using the scanning electron microscope.' *Proceedings of the Second Annual Scanning Electron Microscope Symposium*, 197-203.

KORAN, Z. and CÔTÉ, W. A. 1964. 'Ultrastructure of tyloses and a theory of their growth mechanism.' *International Association of Wood Anatomists News Bulletin No. 2.*

KORAN, Z. and CÔTÉ, W. A. 1965. 'The Ultrastructure of tyloses.' In *Cellular Ultrastructure of Woody Plants*, ed W. A. Côté, 319–333. Syracuse University Press, New York.

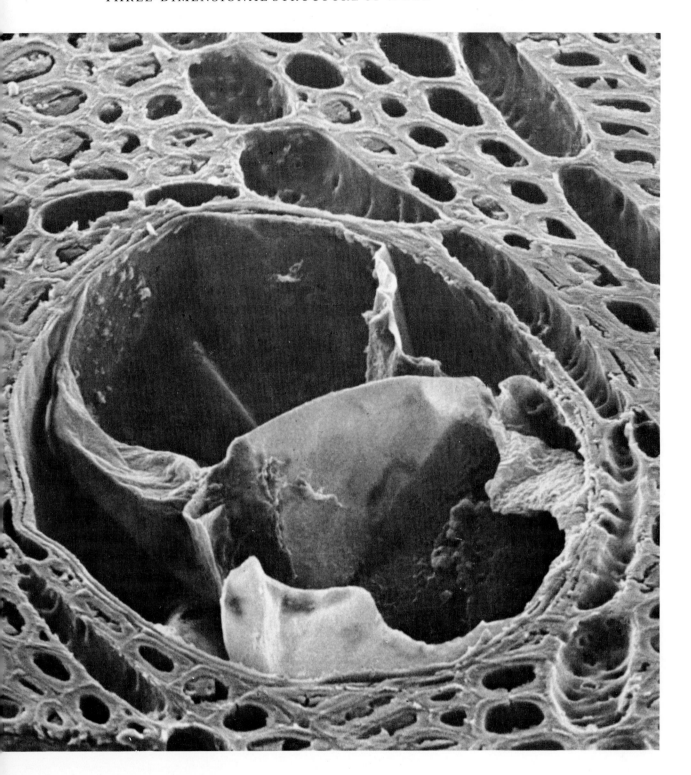

Figure 47. *Tyloses in a vessel of northern rata* (Metrosideros robusta) *seen in transverse section.* (× 1150)

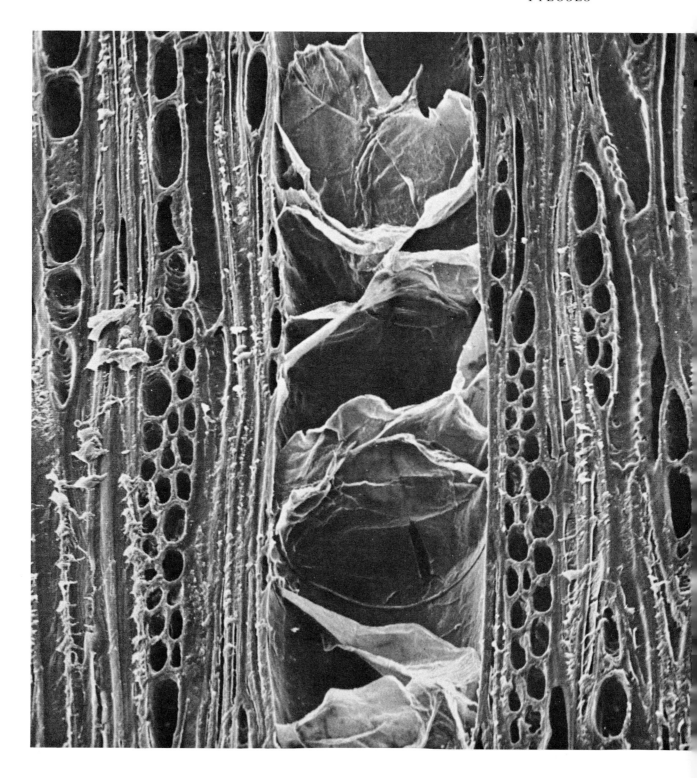

Figure 48. *Tangential longitudinal section of northern rata* (Metrosideros robusta) *showing the thin walled tyloses filling the vessel.* (× 575)

Gymnosperm wood

The secondary xylem of the gymnosperms, commonly called softwood, is simpler and more homogeneous than that of the angiosperms. The main differences between the two types of wood are the presence of vessels in the angiosperms and their absence in the gymnosperms (with the exception of one small order of plants) and the relatively small amount of axial parenchyma found in the gymnosperms. The axial system of the gymnospermous wood is composed almost entirely of tracheids (fig. 49) though small amounts of axial parenchyma are found in some species. The tracheids are long cells with several ray contacts and their ends overlapping each other. The cells are pitted in an opposite or alternate pattern, principally on their radial walls, and these pits in members of the Coniferales and some other genera are of the conifer bordered pit type where the pit membranes possess tori. The pits are normally concentrated at the ends of the tracheids and on those portions of the walls adjacent to rays (fig. 50). Resin canals occur in a number of gymnosperms. Growth rings are usually prominent, the latewood tracheids developing thicker walls with reduced bordered pits.

The ray system conducts materials horizontally across the stem. It contains ray parenchyma and ray tracheids, the latter being distinguished by their thicker walls, bordered pits and lack of a living protoplast at maturity. In the majority of gymnosperms the rays are uniseriate. Horizontal resin canals may occur.

Although the tracheids fulfil the double function of support and conduction, and the wood lacks the apparent sophistication of vessels with open perforations and other modifications found in the angiosperms, the xylem of gymnosperms appears to be a highly efficient system. The giant redwoods of North America (*Sequioa sempervirens* D. Don.), for example, grow to more than 300 feet in height. The trunks of these giant trees, therefore, act as a water-conducting pathway for a remarkable distance when it is remembered how comparatively small the cells of the xylem are. These trunks also hold huge weights of branches and foliage against severe wind loads.

Further Reading

ESAU, K. *Plant Anatomy*. Wiley, New York.
JANE, F. W. 1970. *The Structure of Wood*. Black, London.
PANSHIN, A. J. and de ZEEUW, C. (1970). *Textbook of Wood Technology*. McGraw-Hill, New York.

GYMNOSPERM WOOD

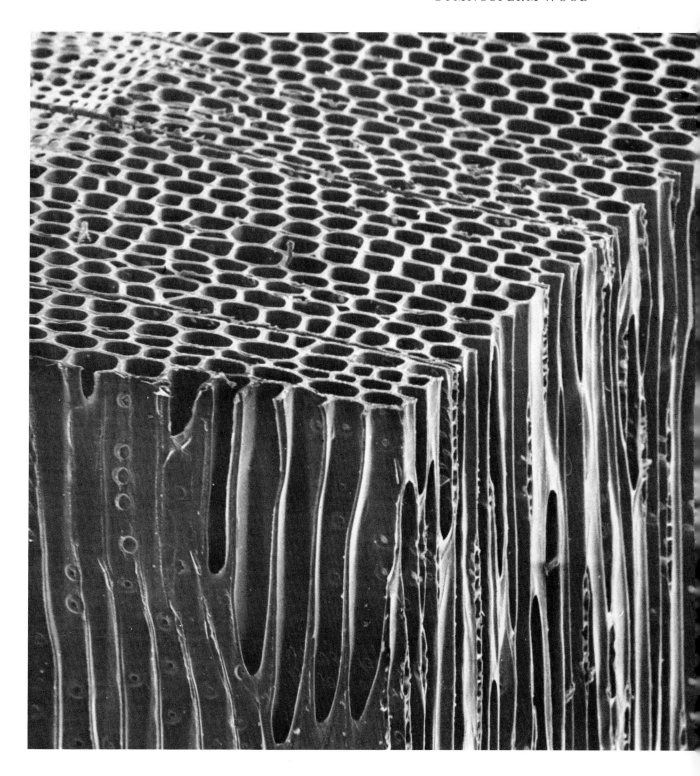

Figure 49. *A three dimensional view of* Pinus radiata *wood. From the transverse cut it can be seen that the bulk of the axial system is built up of tracheids. A growth ring boundary is visible at the top left. Rays, as well as tracheids, can be seen in the tangential longitudinal surface on the right of the photograph. These rays are predominantly uniseriate.* (× 230)

Figure 50. *A tangential longitudinal surface of Douglas fir* (Pseudotsuga menziesii). *Helical thickening occurs on the tracheid walls and the conifer bordered pits are a feature of the radial walls. Uniseriate rays and a multiseriate ray with a resin canal can be seen.* (× 400)

Angiosperm wood

Although some monocotyledonous plants (e.g. the palms) do show some secondary growth, they do not possess a complete cylinder of vascular cambium and hence their stems do not contain wood in the usual sense of the word. Many dicotyledonous angiosperms, however, produce considerable amounts of secondary xylem. This wood, generally referred to as hardwood, is considerably more complex than that found in the conifers since it contains a greater variety of cell types (figs. 51 to 55). Whereas in the softwood the axial system is built up almost entirely of tracheids, hardwoods have evolved two different cell types for conduction and support. Vessels, built up of individual vessel members joined end to end, provide a very efficient pathway for the ascent of sap up the tree. These vessels are variously distributed within the growth rings.

The function of mechanical support in angiosperm wood is carried out by the wood fibres. These extremely long cells with thick walls, arranged in various groupings within the growth rings, act as reinforcing members within the axial system. Libriform fibres are longer than fibre tracheids and have only simple pits. Fibre tracheids are shorter and have reduced bordered pits. In some plants the fibres can occupy more than half the volume of the wood and form the bulk of the fibrous mass obtained when the wood is pulped. Axial parenchyma cells are more abundant in hardwoods than in softwoods where they occur only rarely.

It is generally accepted that vessel members and fibres in hardwoods have evolved from tracheids, so that two separate cell types have become specialized to fulfil the two functions undertaken by the tracheid in other plants. Primitive dicotyledonous woods possess long vessel members with very oblique scalariform perforation plates while more specialized woods contain short wide vessel members with transverse simple perforation plates.

Rays occupy a higher proportion of the wood volume in angiosperms than in gymnosperms. They are usually multiseriate and may contain both procumbent and upright cells.

Further Reading

CARLQUIST, S. 1961. *Comparative Plant Anatomy*. Holt, Rinehart and Winston, New York.
ESAU, K. 1960. *The Anatomy of Seed Plants*. Wiley, New York.
JANE, F. W. 1970. *The Structure of Wood*. Black, London.
PANSHIN, A. J. and de ZEEUW, C. (1970). *Textbook of Wood Technology*. McGraw-Hill, New York.

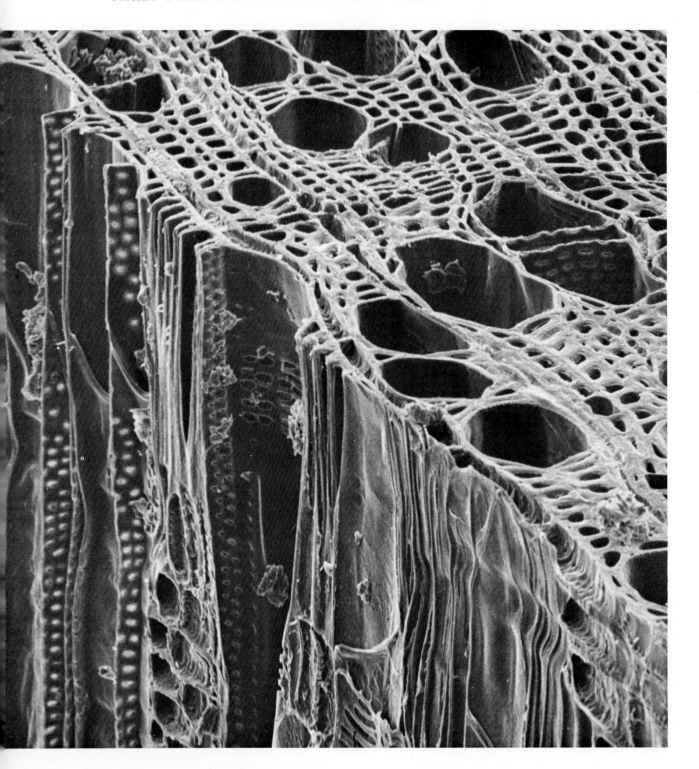

Figure 51. *Vessels, fibres and ray parenchyma cells can be seen in this photograph of red beech* (Nothofagus fusca) *wood. Simple perforation plates between individual vessel members can be seen in three of the vessels in the radial longitudinal surface and a larger vessel shows cross-field pitting where the ray which emerges at the lower right passes behind it.* (× 375)

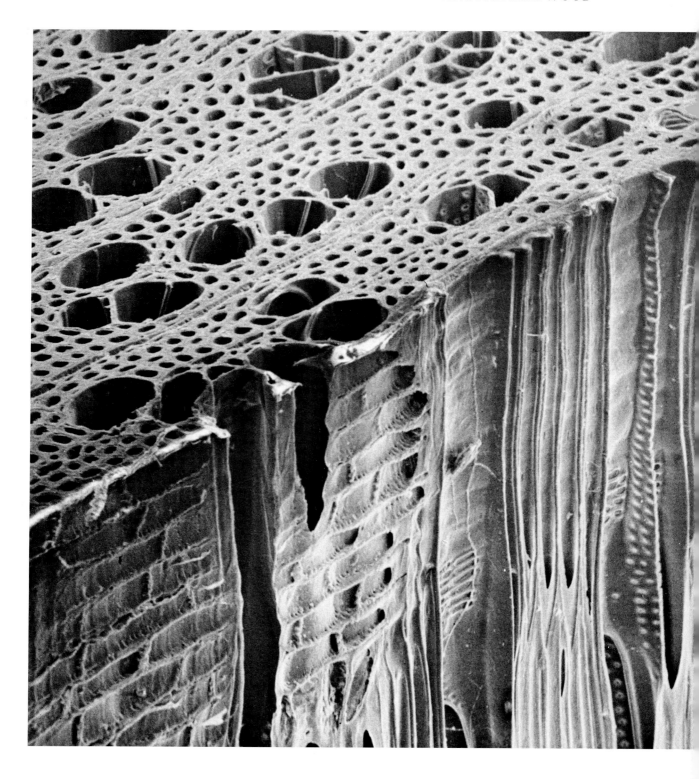

Figure 52. *Simple perforation plates between vessel members can be seen in most of the vessels in the transverse plane of this block of mountain beech* (Nothofagus solandri *var.* cliffortioides *Hook. f.) In the radial longitudinal plane a ray has been cut slightly obliquely.* (× 400)

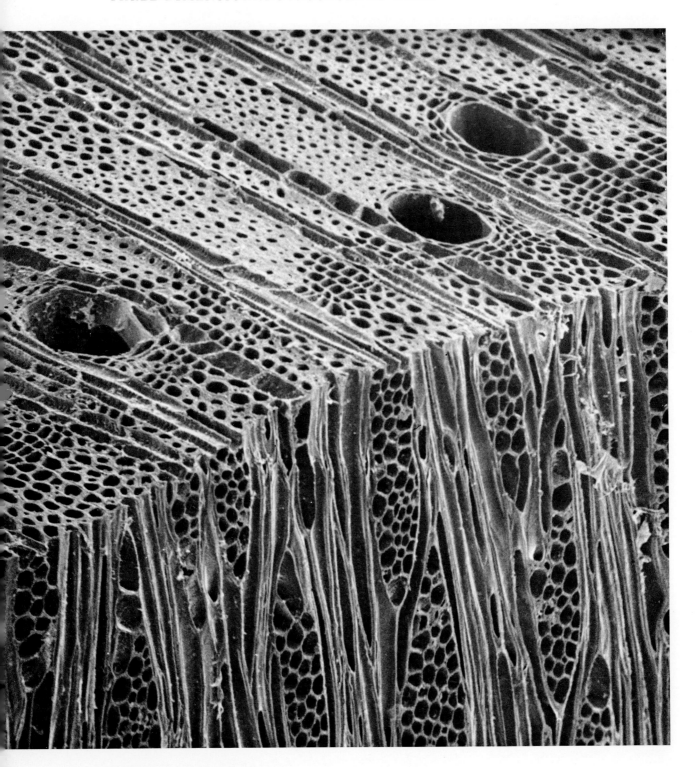

Figure 53. *The numerous multiseriate rays in the tangential cut of this block of* Beilschmiedia tawa *also show in the transverse plane. Vessels and fibres are also illustrated.* (× 190)

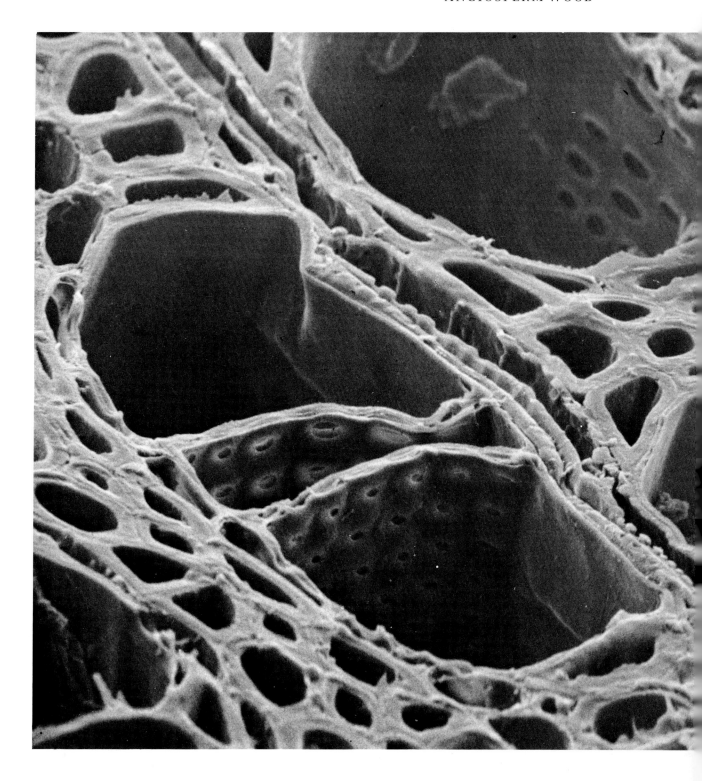

Figure 54. *Vessel to vessel pitting can be clearly seen between the vessels in the centre of this photograph of red beech* (Nothofagus fusca) *wood. A ray runs from top left to bottom right.* (× 1450)

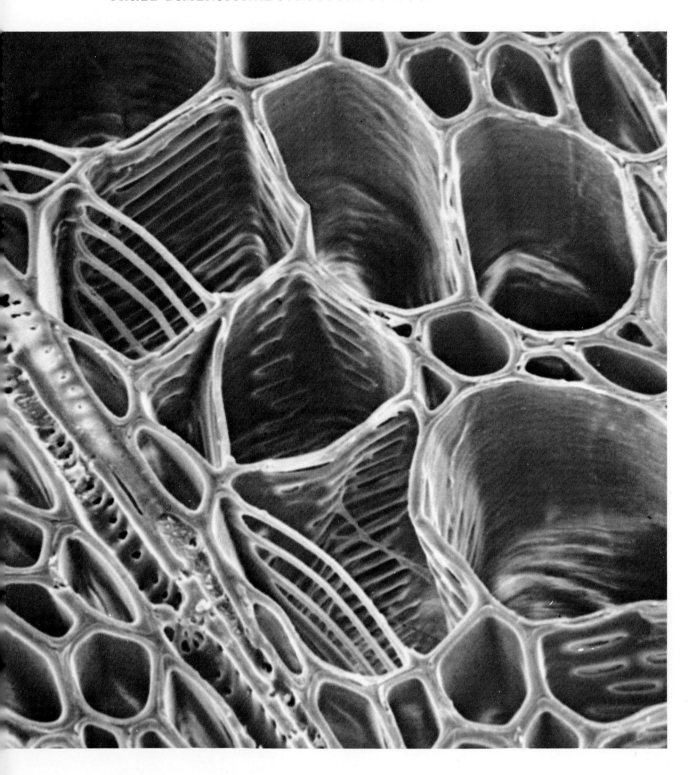

Figure 55. *Scalariform perforations grade into scalariform pitting in the vessel members of this* Magnolia *sp. A ray runs diagonally across the left of the photograph.* (× 1200)

Reaction wood

Reaction wood is modified wood found in leaning branches and trunks of trees. It is accompanied by a marked eccentric growth of the stem. This eccentricity is directed to the lower side of the branch in the gymnosperms where the modified wood is termed *compression* wood. In the angiosperms the eccentricity containing the modified wood is directed to the upper side and is therefore termed *tension* wood.

(a) *Compression wood*

Gymnosperm compression wood is characterized by wider growth rings with less distinct ring boundaries. The tracheids are rounded in outline and show intercellular spaces when seen in transverse view (figs. 56 and 57). When viewed longitudinally, long helical splits are seen in the cell wall with finer helical striations on the wall surface (figs. 58 and 59). In compression wood tracheids the S3 layer of the secondary wall is usually absent, or at least very poorly developed. The helical checks are a feature of the thick S2 layer only and do not traverse the S1 layer or the primary wall. Warts are more sparsely distributed than in the normal wood of the same species and are often confined to the grooves.

Compression wood is known to be a serious defect in timber. Although it appears to be a denser than normal wood it is structurally weaker. It has a lower modulus of elasticity and a lower tensile and impact strength than normal wood. Timber containing compression wood also has a greater tendency to warp and split due to the higher longitudinal shrinkage of compression wood tracheids on drying. Its pulping properties are also inferior.

(b) *Tension wood*

The eccentricity developed to the upper side of leaning branches of angiosperms consists of tension wood. Anatomically the wood is denser than normal wood, the vessels are more sparsely distributed and are smaller in diameter. Thick walled *gelatinous fibres* are a prominent feature of reaction wood (fig. 60). These are cells which contain a gelatinous or G-layer inside the S3 layer, replacing the S3, or replacing both the S3 and S2 layers. The G-layer is an unlignified zone of low angle microfibrils, easily separated from the rest of the secondary wall (fig. 61).

Although not as high as that of compression wood, the longitudinal shrinkage of tension wood is nevertheless much higher than that of normal hardwood. Although it has a high tensile strength when air dried, tension wood is a serious problem in timber since it may collapse, owing to excessive and uneven shrinkage.

Further Reading

CÔTÉ, W. A. and DAY, A. C. 1965 'Anatomy and ultrastructure of reaction wood. In *Cellular Ultrastructure of Woody Plants*, ed W. A. Côté, 391-418. Syracuse University Press, New York.

DADSWELL, H. E. and WARDROP, A. B. 1955. 'The structure and properties of tension wood.' *Holzforschung*, **9**, 97-104.

ROBARDS, A. W. 1969. 'The effect of gravity on the formation of wood.' *Science Progress* (*Oxf.*) 57, 513–532.

SCURFIELD, G. and SILVA, S. R. 1969. 'The structure of reaction wood as indicated by scanning electron microscopy.' *Australian Journal of Botany*, **17**, 391-402.

WARDROP, A. B. 1965. 'The reaction anatomy of arborescent angiosperms.' In *The Formation of Wood in Forest Trees*, by A. H. Zimmermann, 405-456. Academic Press, New York.

WHITE, D. J. B. 1965. 'The anatomy of reaction tissues in plants. In *Viewpoints in Biology*, IV, ed J. D. Carthy and C. L. Duddington, 54–82. Butterworth, London.

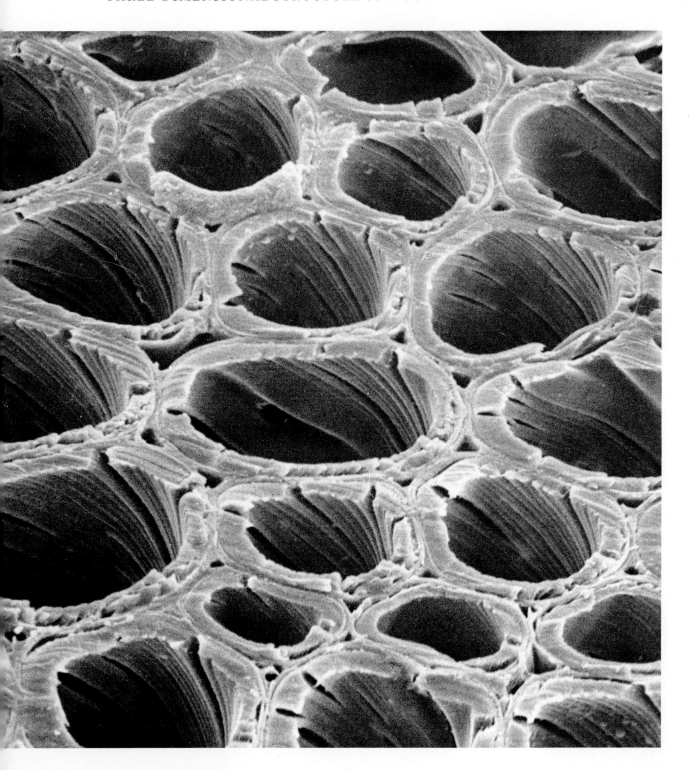

Figure 56. *Transverse section through compression wood in* Pinus radiata. *Note the helical checks and finer striations in the cell walls.* (× 1650)

REACTION WOOD

Figure 57. *Compression wood tracheids in* Pinus radiata.
Note the intercellular spaces. (× 5500)

THREE-DIMENSIONAL STRUCTURE OF WOOD

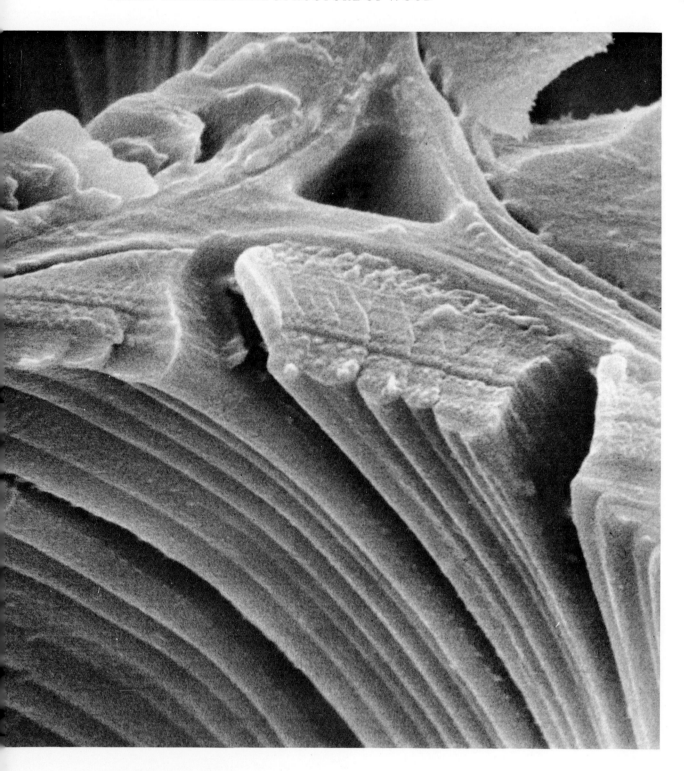

Figure 58. *A close up view of the wall structure of compression wood tracheids in* Pinus radiata. *Notice how the S2 wall appears to consist of 'ribbons' of cell wall material separated at intervals by the helical checks. These splits do not traverse the primary wall or the S1 layer.* (× 10,000)

Figure 59. *Surface view of the helical checks in a compression wood tracheid of* Pinus radiata. (× 13,000)

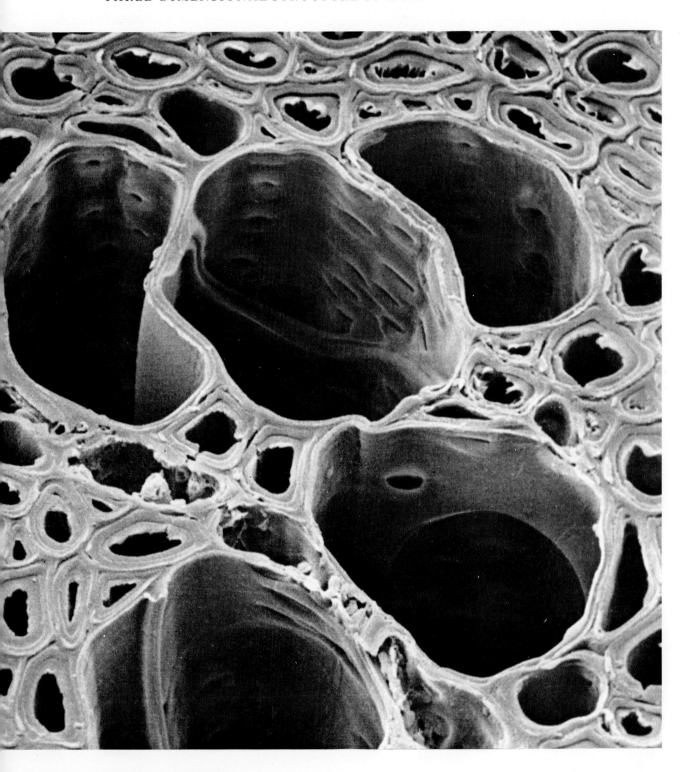

Figure 60. *Gelatinous fibres surround the vessels in this transverse section of beech* (Fagus sylvatica). (× 2000)

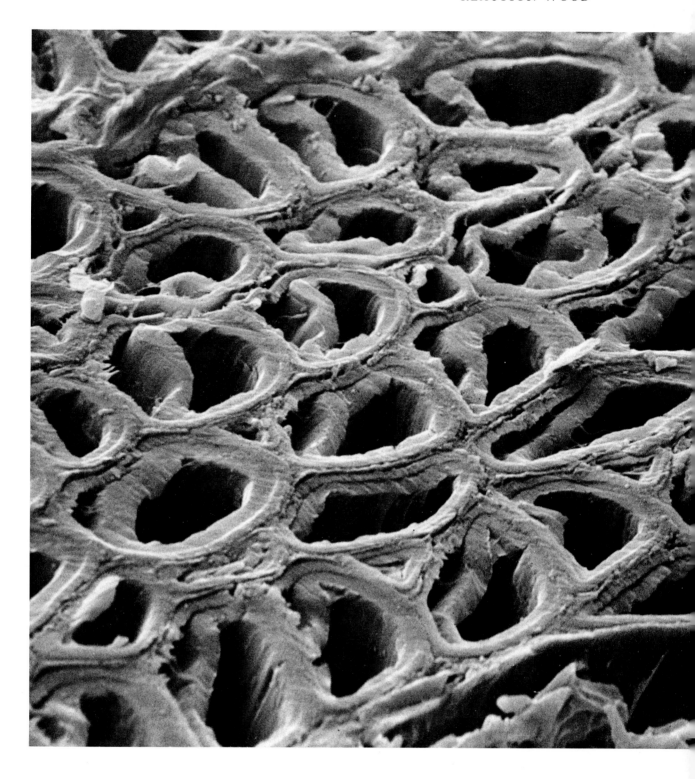

Figure 61. *In this close up view of gelatinous fibres in willow* (Salix *sp.*) *the G-layer has separated from the rest of the secondary wall.* (× 4000)

Index

(*Illustrations are indicated by page numbers in bold type*)

Agathis australis, 22, **24**
Angiosperm wood, 67
Araucaria, 22
Aspirated pit, 13, **19**
Axial parenchyma, 39, 46, **47**, 67
Beech, european, 10, 11, **41**, 78
    mountain, **69**
    red, 25, **27**, 68, **71**
Beilschmedia tawa, 26, **70**
Cellulose microfibrils, 8, **18**
Compression wood, 73, **74-77**
Cross-field pitting, 31, 39, **45**, 68
Diffuse porous vessel distribution, 51, **52**
Douglas-fir, 29, **30**, 42, 59. **60**, 66
Ducts – see resin canals, 58
Elaeocarpus dentatus, 44, **55**
Eucalyptus delegatensis, 20, **34** 54
Fagus sylvatica, 10, 11, **41**, 78
Fibres, 12, **26**, 67, **68**
Fibre tracheids, 22, 67
Gelatinous fibres, 73, **78**, **79**
Griselinia littoralis, 36
Growth rings, 49, **50**, **52**, **53**, **54**, **65**
Gymnosperm wood, 17, 64, **65**
Hardwood – see Angiosperm wood, 67
Helical thickening, 16, 28, **29-32**, 42, **55**
Hemicellulose, 8
Hinau, 44, **55**
Hoheria populnea, 32, **43**, **47**, **48**
Kauri, 22, **24**
Knightia excelsa, 35, 51, **56**, **57**
Lacebark, 32, **43**, **47**, **48**
Laurelia novae-zelandiae, frontispiece, 7, **23**
Libriform fibres, 22, 67
Lignin, 8
Lumen, 8, 12
Magnolia, **72**
Margo, 12, 13, **16**, **18**
Metrosideros robusta, 21, **62**, **63**
Microfibrils, 8, 12–13, **18**

Microscopes, 6
Middle lamella, 8, **8**, **9**, 12
New Zealand honeysuckle, 35, 51, **56**, **57**
Nikau palm, 37
Northern rata, 21, **62**, **63**
Nothofagus fusca, 25, **27**, 68, **71**
    solandri, **69**
Parenchyma, 22, 39
    axial, 39, 46, **47**, **48**, 67,
    ray, 39, **40**, **41**, **43**, 68
Parenchyma strands, 46, **47**
Perforations, 12
Perforation plates, 33, **34-38**
    multiple, 33, **36**, **37**
    reticulate, 33, **37**
    scalariform, 33, **36**, **72**
    simple, 33, **34**, **35**, **38**, **68**, **69**
Pinus radiata, 9, **17**, **19**, **40**, **50**, 65, **74-77**
Pit aperture, 12, **17**
Pit canal, 12,
Pit cavity, 12
Pit chamber, 12, **17**, **19**
Pit membrane, 12, **14**, **15**, 35
Pit-pairs, 12, **14**
    bordered, 12
    crossed, 13
    half bordered, 12
    reduced bordered, 26, **36**
    simple, 12, 37, 43
Pits,
    aspirated, 13, **19**
    blind, 12
    bordered, 12, **17**, **42**
    branched, 13
    reduced bordered, 12, **25**, **26**
    scalariform, 23
    simple, 12, 13, **43**, **47**, **48**
    vestured, 13, **20**, **21**
Pitting,
    alternate, 22, **24**
    cross-field, 31, 39, **45**, 68
    irregular, 22
    opposite, 22
    scalariform, 22, **23**, **72**
    unilateral compound, 12, **16**
Plagianthus, 51

Populus, 52
    nigra, **45**
Pore, 51
    distribution, 51, **54**, **55**
    see also vessels, 33, **34-38**, **57**, **68**, **69**
Primary wall, 8, **9**, 12
Pseudopanax arboreum, 14, **16**, 31
Pseudotsuga menziesii, 29, **30**, 42, 59, **60**, 66
Rays, 39, **40-44**, 67
    multiseriate, 39, **41**, **42**, **60**, **66**, **70**
    uniseriate, 39, **40**, **42**, **65**, **66**
Ray parenchyma, 39, **40**, **43**, **64**, 68
Ray tracheids, 64
Reaction wood, 73
Red beech, 25, **27**, 68, **71**
Redwood – see *Sequoia sempervirens*, 64
Resin canals, 58
    axial, 58, **59**
    horizontal, 42, 58, **60**
Rhopalostylis sapida, 37
Ring porous vessel distribution, 51, **53**
Salix, **79**
Sclerosis, 61
Secondary wall, 8, **9**, 12, **28**
Sequioa sempervirens, 64
Softwood – see Gymnosperm wood, 17, 64, **65**
Tension wood, 73, **78**, **79**
Torus, 12, 13, **18**, **19**
Tracheids, 9, 12, **17**, **29**, 65
Tyloses, 61, **62**, **63**
Ulmus, 38
    glabra, **53**
Vessels, 33, **34-38**, **57**, **68**
    distribution, 51, **52-57**
Vessel members, 10, 11, 13, **27**, **34-38**, **45**, **71**, **72**
Vestures, 13, **20**, **21**
Xylem, 8, 12
Warty layer, 8, **9-11**, 13, **20**, **21**, 73
Willow, **79**